THE FORGOTTEN GRIST and FLOUR MILLS OF WALLOWA COUNTY, OREGON

A history of the eight grist or flour mills that have operated in Wallowa County, Oregon.

The first mill was built in Joseph in 1882
Wallowa was the last to dismantle their mill in 1960.

Courtesy Gene Hayes

Wallowa Mercantile Company Roller Mill

Best Wishes
Irene Barklow

IRENE LOCKE BARKLOW

Enchantments Publishing of Oregon
504 Donald Street, Wallowa, Oregon 97828

Edited by Jack Evans

ISBN 09618185-5-7 (PB)

Library of Congress Cataloging-in-Publication Data

Barklow, Irene Locke
The Forgotten Grist and Flour Mills in Wallowa County, Oregon: a history of the eight mills that have been in this area.

Includes index.
1. Mills - Wallowa County, Oregon 2. Gristmills - Wallowa County, Oregon 3. Wallowa County, Oregon - History I. Title

Front Cover: Troy Roller Milling Company, Troy, Oregon
Courtesy *Enterprise Record Chieftain May 14, 1914*

TABLE OF CONTENTS

Dedicated to my family:
Kenneth Barklow
Christine, Roark and Margo Schwanenberg,
Christopher and Deena Tsiatsos.

ACKNOWLEDGEMENTS

When I started researching material for this book, I realized a book about the flour mills should have been written years ago. Very few people are alive who have knowledge about the mills.

My mentor, Jack Evans, was the person that encouraged me to write my first book *The Old and The New*. Again he came to my rescue and supplied the material on the working parts of the mills and other important information, plus editing this book. I say thank you, Jack.

As with my other books the crew at the *Wallowa County Chieftain* office was very helpful. The back editions of different papers were made available. All the assistance I needed was given to me. Thanks, again *Wallowa County Chieftain* people.

I want to thank all the people who have helped me research this story: Marilyn Hulse, Francis Crow, Phil Bowman, Marshell Thiesen, Arleigh Isley, Lois Blankenship, Mary Burrows and others already mentioned. A special thanks goes to the Community Bank for use of their copy machine. I also appreciate the use of the paper cutter at the Enterprise Stationery store. Thanks to Mr. Tony Flagg from the Pendleton Flour Mills, for the loan of his book. Jo Ann Snead, Bobbie Long, Beulah Wynans, Moonlight Graphics and Bernice McBath supplied their old flour sacks to be copied. Jo Ann Snead has a quilting business called "Mouse Paw Quiltworks". Ruth Dressler gave me flour sack fabrics. Jennifer Hobbs came to my rescue and helped me immensely with the operation of my new computer and helped me use the editing tools.

I bought my new computer from Henderson Friendly Computers. Leonard Henderson and Bill Peach were both jewels to work with. Leonard scanned and set the photographs in this book. Thank you, Leonard. In addition, I appreciate the assistance from William Shaver and Marlene Zollman.

The old plat maps of Enterprise, Joseph and Wallowa, courtesy of the Wallowa County Library and Eastern Oregon University were special finds. These maps listed the type of machinery that was used in each mill, besides showing the water spillways, overflows and other things I had been trying to find the answer to. Wow, what a happy day when I found the microfilm with the plat maps illustrated. The Sanborn Company originally published these maps for use in connection with fire insurance. Each building in the community was shown, with its floor plan, and the maps were color coded to indicate type of construction. Details were given for the mills such as source of power, kind of lighting, fire protection available, and so on.

My husband Kenneth Barklow and our daughter Christine Schwanenberg were always there when I needed assistance. I called for help often. I am blessed by their thoughtfulness.

Wallowa County is very fortunate to have so many knowledgeable volunteers at the Wallowa County Museum. I appreciate all the friendly assistance. The County and City Libraries have been very informative and thoughtful in their help. Also the Wallowa County Clerk's office is always supportive.

INTRODUCTION

THE FORGOTTEN GRIST and FLOUR MILLS OF WALLOWA COUNTY, OREGON

The building of the flour mills in the county was real progress and a very important part of the prosperity that followed. As additional land was homesteaded and more farmers planted grain, gristmills were a necessity. Grain was exported to Asia from the valley mills. All the mills seemed to flourish until after World War I. Then the government stopped buying flour, which hit them hard. Also, better transportation into the area made it possible for products produced elsewhere to reach the Wallowas. Progress and time change everything, but the flour mills were essential for the development of the county. Four of the mills in the valley furnished the first electricity to each town.

Farmers started purchasing their own feed mills as early as 1907. The *Wallowa Chieftain* stated, October 10, 1907 "James Kooch has purchased a gasoline engine and feed mill, to be used at his ranch. Already grunts of delight are emanating from the numerous pig pens upon his place."

There was a grist (flour) mill in each of the following locations: On Spring Creek near the present Enterprise fish hatchery (the date Mr. Edwin Parrish started this mill is not known); Mr. Hiram Williams placed his mill where Parsnip Creek enters the Wallowa river, a couple miles north of Lostine, (1885); also, a mill at Joseph is said to have operated as early as 1882. The more important mills came later. The one in Enterprise was built in 1888, Wallowa in 1894, Lostine in 1895, and Flora in 1908. Troy, the last on the list, was built in 1914.

Burr or granite millstones were used in the earliest mills. Later they converted to roller-mills. This was a much easier and faster means of grinding.

Four of these mills burned. Lostine rebuilt; Joseph had two fires, one in 1889 and 1921; Enterprise did not rebuild their mill after it burned in 1926.

One reference was found in the *Wallowa Chieftain*, December 22, 1897 in the Prairie Creek items. "C.W. Meek has repaired his water wheel and is again ready to smoothe (sic) your lumber or crush your barley." Evidently Meek just crushed grain and didn't grind for flour.

The following information was found mainly in the newspapers. Most of the stories were written at the time the mills burned, changed hands, or were torn down. Almost everyone who knew about the mills has already departed this world.

Roy Carper remembers when George Courtney over on Eden Bench had a grain chopper. They would pour grain in a hopper. Under the hopper was a grinding stone. One horse was hooked up to go around, around and around the hopper. After the grain was chopped or ground it was fed to the animals. They did not make flour at this location. An old grinding stone has been found in the Promise area, at the old Bloom homestead. It probably operated in the same manner as the one in the Eden area.

Chapter I

THE FORGOTTEN GRIST and FLOUR MILLS OF WALLOWA COUNTY, OREGON

The visual encounter of the large impressive flour mills inspired me to explore their histories. I especially enjoyed the picture of the Troy Mill, showing the millrace.

The following article in the *Enterprise Record Chieftain* December 31, 1925 tells about one of the first flour mills in Wallowa County.

FIRST ROLLER MILL IN WALLOWA VALLEY

"The first roller mill in Wallowa Valley, and the second mill of any kind, was built by Hiram N. Williams at the mouth of Parsnip creek in 1885, the year after the railroad was built to La Grande, and it is still in daily operation. (An article on the flour mill at Joseph states that it was built in 1882). Mr. Williams has used it continuously since he put it in and may be seen any day chopping grain to feed to his stock. The only repair bill ever paid was $21 for one new roller. One mill was built a year or so before this; it was the old burr mill put in by Edward Parrish, on Spring creek, just south of Enterprise. . .

HIRAM WILLIAMS MILL

Hiram Williams Mill *Enterprise Record Chieftain*, July 7, 1932

"Hiram N. Williams, a pioneer of 1879, passed away Thursday night June 30, 1932 at his home at the mouth of Parsnip creek. He had lived there since July 26, 1879, when he and his brother, George, shifted the box off their wagon and made camp for the night, at the end of their long overland drive from Illinois. . ."For several years Mr. Williams' health had been failing, with age, but he was ill only about five months, with an affection of the heart. . .

"Mr. Williams and his brother, George, left Missouri April 14, in a wagon, with horses and some colts. They worked four horses and the others followed or were led. At night all the horses were turned out to graze, and grass was none too plentiful and the stock became worn and thin. However, the travelers kept moving except that they stopped three days near Boise, Idaho, and the men worked to get some much needed cash.

"At Fort Steele they fell in with "Bud" Blankenship and Oscar Beidler, who came from the same neighborhood in Missouri, but had started west independently of the Williams brothers. The two outfits came the rest of the way together. The Williams brothers had intended going to Puget Sound but stopped and turned aside into Wallowa valley because Blankenship and Beidler were coming here, and the travelers were out of funds.

"The brothers ended their long trek at the mouth of Parsnip creek, driving their wagon on a low bench above the stream and removing the box and making camp, July 26, 1879. They took up the land on which they had settled and it remained their home ever since.

"Mrs. Williams and two small children remained behind in Missouri until fall and then went by railroad to San Francisco, to Portland by ocean steamer and to The Dalles by river boat, arriving late in December. So great was the pressure of travel to the interior that her turn to come to Eastern Oregon by stage was a month away. However, the illness of other passengers gave her an opportunity to take their turn after only two weeks of waiting, and she came thru to Summerville.

"Mr. Williams met her there, driving out in a wagon. The winter was mild and there was no snow, but so much rain had fallen that the improvised roads were deep in mud. Mrs. Williams reached her new home where she was to spend her remaining days, early in January 1880.

"Mrs. Williams passed away at the family home July 25, 1926. There are four children, J.R. Williams, Mrs. Lulu B. Read, Mrs. Cora Fleener and Mrs. Stella Hook, all living in the valley. There are 13 grandchildren. Four children preceded their parents in death; two passed away when very young in the east and Daisy and Clarence in Wallowa valley. George Williams also survives his brother with whom he had been a partner since their youth in the east.

"The old mill still operated by George Williams is an object of interest to those who know its history, as it gives further evidence of the brothers' stability and persistence . . ."

"Hiram Williams passed away June 30, 1932. George Williams, Hiram's brother operated the mill until his death November 6, 1932." (Florence Read Workman.)

When the estate was settled, the land was divided between the Williams children. Mrs. James (Jim) (Lulabelle or Lula) Read inherited the middle section of land. Cora Fleener received the upper part where Mr. and Mrs. Duke Lathrop now live. Mrs. Charles (Stella) Hook inherited the mill and land in that area. They ground grain for their own use. The Hook family dismantled the old mill sometime in the 1940's. Bob Williams possessed the other section.

The Williams family peddled their graham flour (flour made from entire kernels of wheat) in the towns of Wallowa and Lostine. My husband, Kenneth Barklow remembers his family driving to the mill to purchase graham flour, which was used for hot cereal. When the cereal was cold, it could be cut into slices and fried. While it was still warm it was buttered and hot sugar syrup poured over the top, making a tasty breakfast.

Little evidence of the old mill can be found today. Nyoda Scott said, "the large pieces of concrete, across Parsnip creek that formed a dam, (shown in the picture) were washed away in the flood of 1975". She believes the mill stood in the vicinity of her house. Duke Lathrop said, "a water wheel was used at this mill".

THE PARRISH GRIST MILL

The date Mr. Edwin Parrish started his mill is not known. We could not find a homestead registered to Mr. Parrish. If Mr. Parrish received an abstract of title before 1887 it would have been registered in Union County. Early records of Union County were lost in a fire. Tom Misandaer showed Arleigh Isley where the old burr and pieces of concrete from the old mill are located today (1999) (No water is running in this area now.) The remains are north of the Fish Hatchery road and east of the present fish hatchery. Silver Creek has changed its course since the time of the mill.

Edward Parrish was born in 1838 and died in 1933. The Parrish children were Amanda (Chenoweth), Elvira (Bailey), Belle (Fosner) and Dora (Renny). Mr. and Mrs. Parrish and their child, Sarah, are buried in the Alder cemetery.

Great Grandchildren living in the Wallowa Valley today are Lois Blankinship, Reatha McCormack and Willis Bailey. They were unable to supply any more information about the Parrish family.

The following article in the *Wallowa Chieftain* December 6, 1900 refers to the Parrish flour mill site. Evidently this is when the electricity for Enterprise was separated from the Enterprise Flour Mill. "F.L. Foster who assisted in installing the electric light plant at Lostine, arrived in town Monday, and on Tuesday started a crew of men at work on the old Parrish (flour) mill property, preparatory to installing the plant in this city by January lst."

Arleigh Isley also showed us the site of the old power plant. It is closer to the fish hatchery than the location of the old mill site. No water is running now at this location. The remains of the old dam are still visible.

JOSEPH FLOUR MILL

Mr. W.H. McCully has been called the father of Wallowa County. It is no surprise to find he was instrumental in starting the first flour mill in the County.

The Illustrated History of Union and Wallowa County, 1902 stated, "the first flour mill in Joseph was started in 1882 by W.H. McCully and W.W. Briggs. This building was destroyed by fire in 1889. It was rebuilt by W.H. McCully and was sold in 1891 to the Joseph Milling Company by whom it is at present operated (1902). The mill has a capacity of fifty barrels per day and has always done an exceptionally good business"

The first item found in the newspaper about a Flour Mill in Joseph was the following ad: "November 5, 1885 - SILVER LAKE MILLS - McCully and Briggs - Best Family Flour - Feed, Chop, and The BEST GRAHAM - In the WORLD constantly on hand for sale or exchange - ONLY FIRST-CLASS MILL - In the Wallowa Valley - Joseph, Oregon." The same ad appeared December 22, 1887.

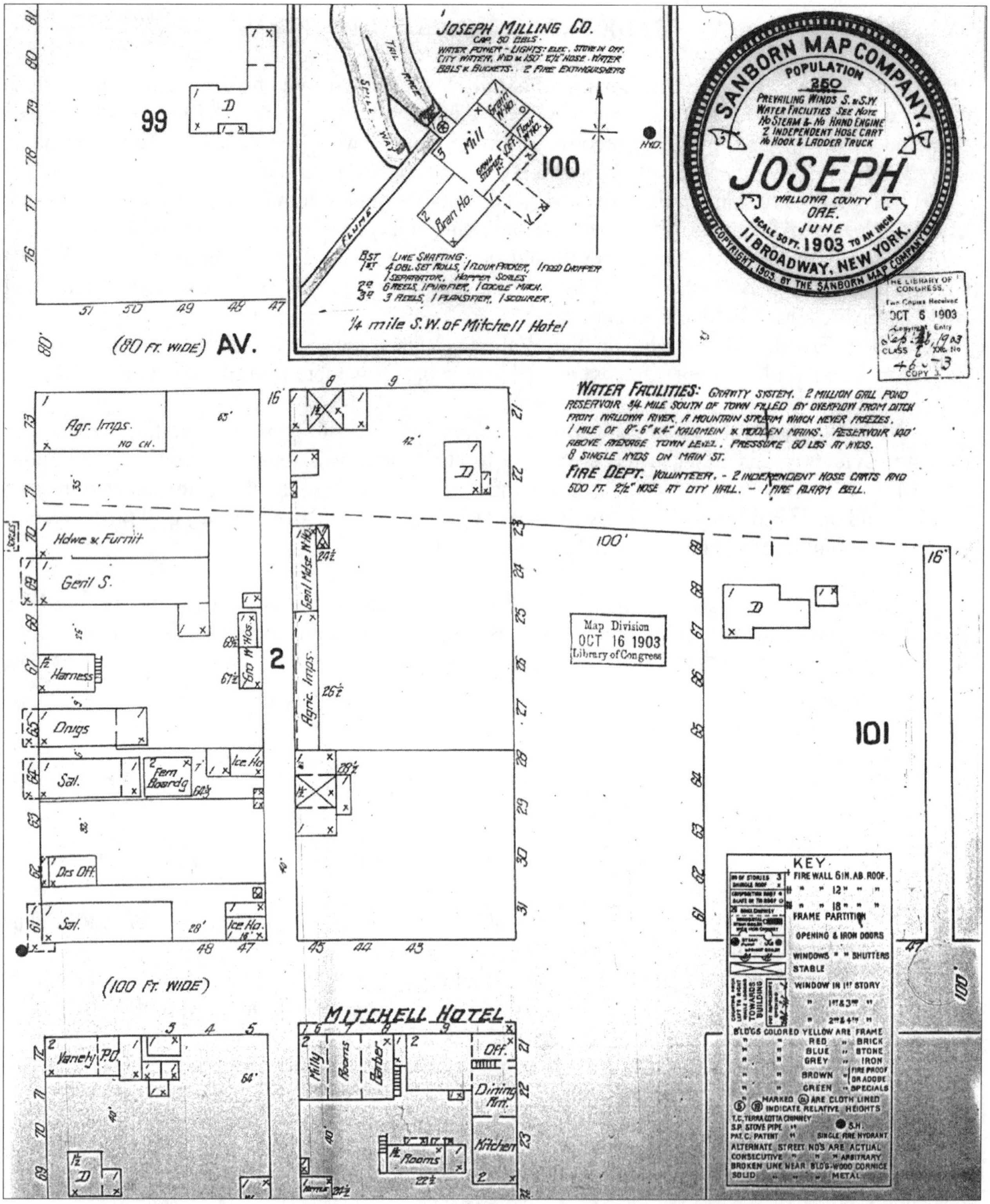

The Joseph Milling Co. inset at the top of the page states, " ¼ mile to the Mitchell Hotel" which was located on 1st and Lake Street. Water for the town was from a mountain stream." Notice other informative information under **Water Facilities and Fire Dept.**

In the *Wallowa Chieftain*, September 25, 1890 it states, "The machinery for the new Silver Lake Mills has nearly all arrived, and is being placed in position as rapidly as possible. Work will probably be farenough [sic] along next week so that wheat can be received, allthough [sic] will likely require three or four weeks yet before all the machinery can be operated." (The first Silver Lake Mill burned in 1889.)

The first record of a deed found was dated December 13, 1894 from F. D. McCully to Joseph Milling Co. On June 25, 1926 the Joseph Milling Co. was transferred to H. A. Bodmer.

The following ad appeared in *The Aurora*, March 30, 1894 and again July 19, 1901.

Joseph Milling Co. . . . Joseph, Oregon.
Manufacturers of FLOUR, FEED AND SELF RISING FLOUR
FULL ROLLER PROCESS.
Chopping done any time.
Highest market price paid for grain.
Flour and Feed sold at the Mill for cash or exchange only.

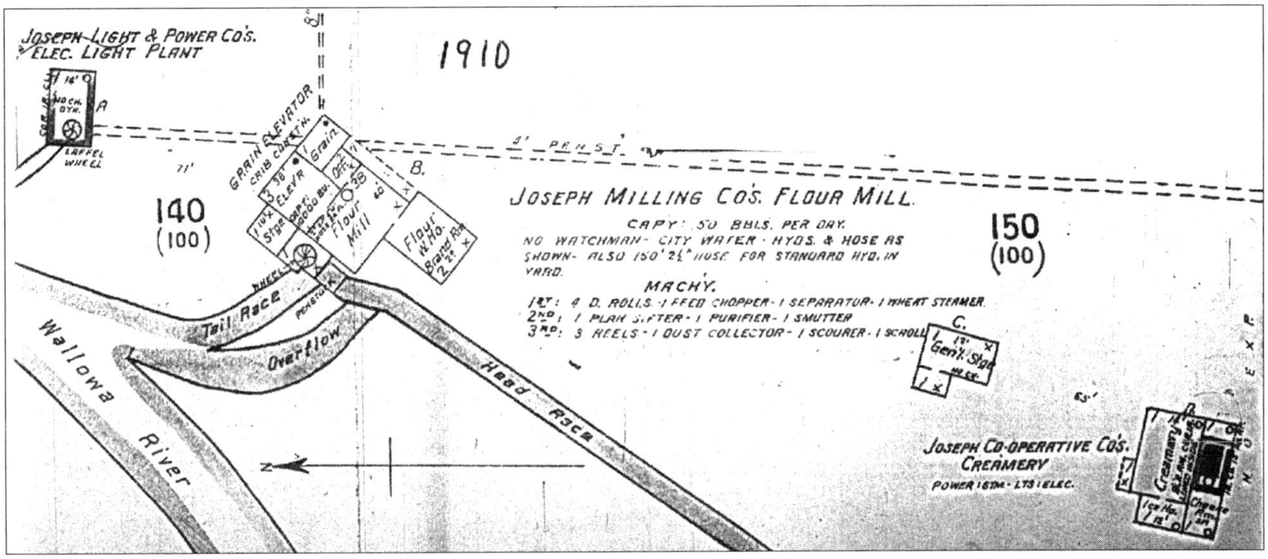

This 1910 map shows where the Joseph Light Plant was located and it is still in the same site today. The mill had been enlarged since 1903. Note the location of the Joseph Co-operative Co.'s Creamery.

The mill was located on the Wallowa River, to the west of the present Pacific Power and Light power station. Mill St. received its name for the flour mill. The millrace can still be seen in the Joseph City Park.

The Joseph Herald, January 1906

"For the past week the Joseph Milling Company have been taking advantage of the fine sleighing and averaged about 8000 pounds of flour daily on their stage to Elgin. Their product, White Satin Flour, is eagerly sought after by buyers for the oriental trade and Oregon Consumers.

This is the only picture found of the **Joseph Flour Mill**

Courtesy *Enterprise Record Chieftain*, June 6, 1918

"For many years the Joseph flour mill has been grinding out feed for man and beast. It was one of the first industries in the town and has remained on the job with few rests. Recently it was enlarged to a capacity of 60 barrels a day. Last fall the mill was run night and day and it has now ground out its allotment, (for the government for World War I) but has been working since on local business."

Enterprise Record Chieftain - January 5, 1922

Mill at Joseph Burned at Night

Building, Wheat and Machinery Destroyed with a Loss of $40,000

"Fire of unknown origin destroyed the flour mill of the Joseph Milling company Saturday night in a spectacular blaze, which was seen for miles in the valley. The mill was a total loss, with its machinery and the wheat and flour stored in it. F. D. McCully the owner of the plant, estimated the loss at $40,000 with insurance of 50 or 60 per cent of its value. About 12,000 bushels of wheat was stored in the mill warehouse, and the ruined grain ran out upon the ground or was carried away in the river in large quantities.

"It is taken for granted that the mill will be rebuilt, . . . The milling company, meanwhile, continues in business, dealing in grain, feed and flour, at its old office on Main Street in Joseph.

"The mill had been running regularly, with S. E. Hambelton in charge. He left his work Saturday at 6 p.m. first going thru (sic) the building to make sure the fire was out in the stove and everything was in perfect condition.

"At about 9 o'clock, families living nearby saw flames bursting from the main door at the south end of the building. The fire seemed to be burning in the basement and was already threatening the end of the structure. The fact that the fire began in this part of the building practically doomed it from the first.

"A two-inch stand pipe was in the building with hose all connected ready for instant use. This was in the south end, near the main door and the office. As the flames enveloped this part of the structure, nobody, could get in to the pipe and hose.

"A line of hose was run from a city hydrant not far away, but the hydrant was frozen. The hose was carried back to another hydrant, two blocks distant, but by this time the structure was in flames, and burning with such intense heat that nobody could get near it.

The night was still and so the embers did not fly far, but the house of St. L. Osborne, not far away on the east, was scorched and was in great danger of destruction. The plant of the Enterprise Electric Company, still closer to the burning mill, became so hot that it could not be occupied. It is of metal construction and so could not burn.

"Men eager to help removed a bale of 1000 grain sacks, and much sacked feed from the porch, and these were carried far enough away so that they seemed to be safe. However the free progress of the fire shortly created such intense heat that these goods were burned after all. The books of the company were kept at its office on Main Street and so were not touched.

"The building of wood, stood on a stone foundation which rose to the first floor. No rapidly turning journals were in the south end of the mill, either in the basement or on the first floor. As Mr. Hambelton inspected the building before leaving it. No clew (sic) of the origin fire is available.

"The first flour mill in the county was erected in 1883 (1882) on the site of the structure burned Saturday night. This first mill burned in 1890 (1889), and the main structure of the new mill was put up two years later. Other additions were made since then.

"The mill was run by water power, and the power right was the oldest in the county, this being the first appropriation for the purpose made from the Wallowa River. Some years ago Mr. McCully formed the Joseph Electric company, which used part of his original water right. When he sold the electric business to the Enterprise Company, the entire water right went to that company, but Mr. McCully received a perpetual right to the use of sufficient electricity to run the mill. However the water power machinery was in the mill and so has been used ever since.

"In 1917 the mill was remodeled and much new machinery installed. Prices of this machinery have advanced a good deal since then, making the loss all the more severe. The mill was an efficient plant with a capacity of 60 barrels of flour a day and served a large territory."

"The first electric plant in the valley was established in 1900 in Joseph by F. D. McCully, banker, miller, and merchant. It consisted of a 30-kilowatt inductor alternator driven from a line shaft in the flour mill. This served the needs of the town after a fashion until 1909, when a modern plant [for that time] was installed, consisting of a 400-hp water wheel directly connected to a 225-kva generator. This was supplied with water through a 4,000 -foot wood stave pipeline three feet in diameter under a 125-foot head. This plant continued to operate until June 1927, when it was practically worn out. The final touch came when a large piece of iron broke loose inside the wheel casing and ripped all the vanes off the water wheel runner. This ended the career of this plant which at one time was the pride of the whole valley." (The above is from the book *The Story of Pacific Power & Light Company and its Pioneer Forebears* by John Dierdorff.)

The Joseph Herald

April 13, 1933 (condensed)

"A flouring mill may be located in Joseph during the coming months. B. F. Locke, owner of the mill at Halfway, was in town talking about moving his mill to Joseph."

"The Joseph flour mill burned down a number of years ago. They made flour called 'White Swan'."

No one built a flour mill in Joseph after the fire in 1921.

ENTERPRISE FLOUR MILL

The earliest record found on the Enterprise flour mill was the following transaction. Robert Stubblefield recorded the first deed to the Island City M. & M. Company July 11, 1887. Sec. 2 Township 2S Range E44 Block 21. This is the location of the Enterprise swimming pool in 1997. The old mill flume is used to drain the pool.

ENTERPRISE MILLING COMPANY

December 22, 1887

"Enterprise Flouring Mill - A Company has been incorporated at Enterprise for the purpose of building a flouring mill. Messrs. Church, Gardner, Aaron Wade and R.W. Bloom are the incorporators (sic), and the capital stock of the company is fixed at seven thousand dollars."

ISLAND CITY MERCANTILE AND MILLING COMPANY

By Jack Evans

This firm was organized by a group of businessmen and landowners, and by the mid-1880's had become the dominant commercial enterprise in Island City. The principal shareholder, William G. Hunter (1841-1907), was a pioneer farmer and rancher in Grande Ronde Valley, and another prominent farmer, J.L. Caviness, headed the local bank. Hunter's imposing residence still stands four miles north of the town.

In addition to a general store, the ICM&M built and operated a local flour mill that served a wide area. The building is no longer a mill but it is still in existence. Expansion of ICM&M included establishment of similar facilities in other communities.

At that time the Wallowa region was still part of Union County, and for some time its residents had campaigned for a county of their own. Isolation and bad roads made more localized government a necessity. In 1886, when it became evident that a division was likely, Robert F. Stubblefield platted a new town – to be called Franklin – on Bennett Flat. When creation of the new county was assured representatives of ICM&M approached local leaders, offering to build and develop a new community in exchange for free land. They were rebuffed at first but the offer was too good to refuse. The Franklin plat was revised to include a courthouse square, the name was changed to Wallowa City, and John Zurcher deeded a block of the choicest land to ICM&M. The company then made good on its promise to establish a store and develop the town, whose name was soon changed to Enterprise.

Two years later, ICM&M created the town of Wallowa – acquiring the land, platting the community, and setting up the general store.

The stores established by ICM&M were not branches but were set up as independent operations. They were not always M&M's and were not always general. The one in La Grande (currently the Eagles lodge) was an outlet for farm machinery and the one in Wallowa was the Wallowa Mercantile. The parent firm, like most community developers, made a profit for its shareholders through capital investment revenue, sale of improved properties, and from revenue generated by stores and mills until they too were sold to those who operated them.

For more information on the early development of Enterprise, see *5200 Thursdays in the Wallowas,* by Lloyd W. Coffman.

The Enterprise M&M's last building, a fine commercial structure with apartments above, was built in 1916. It is located across the street on the north side of the Wallowa County Courthouse.

Note the horse drawn farm implements displayed on the sidewalk of E.M. & M. Co.

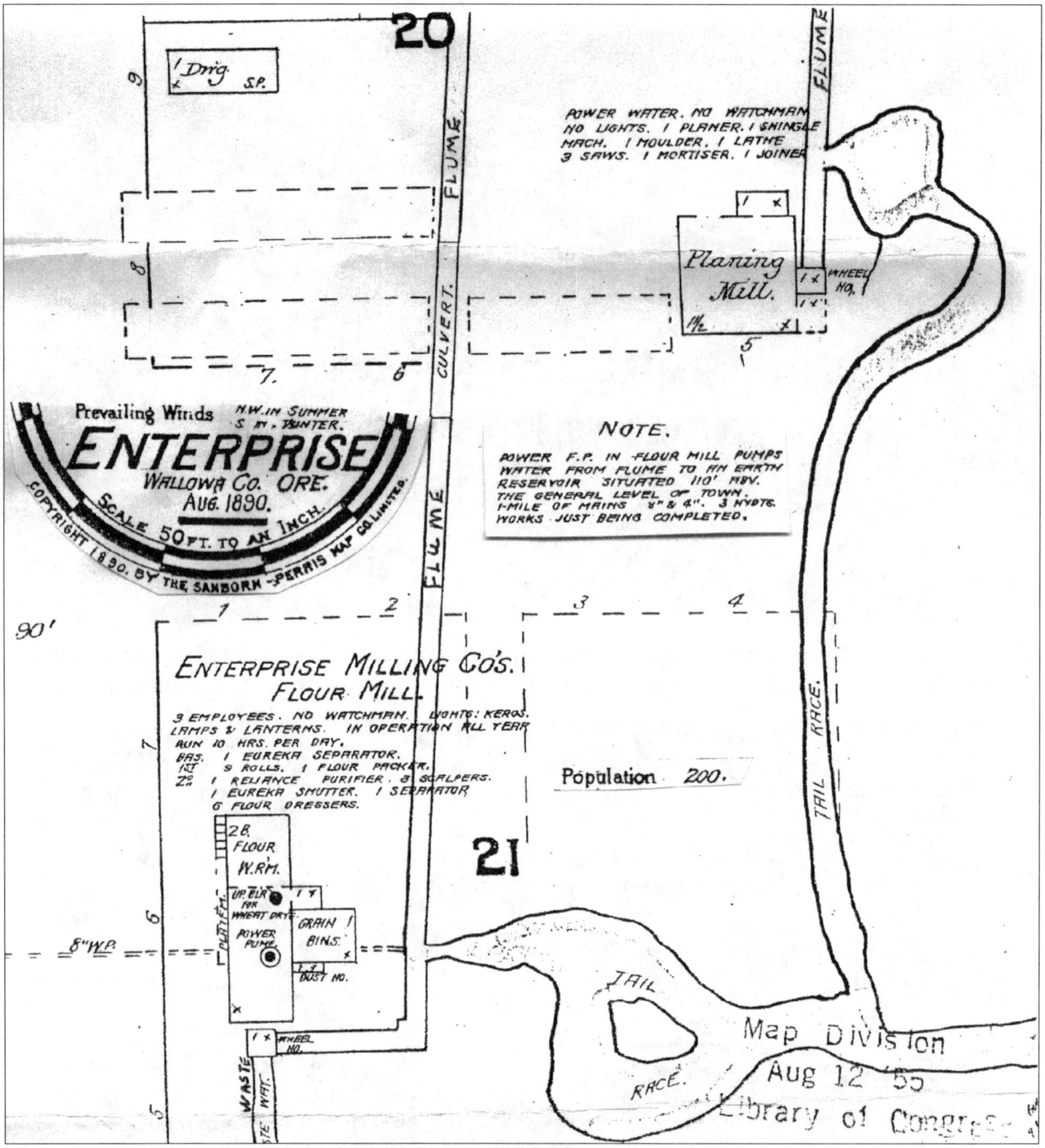

Prairie Creek is located out of sight on the right hand side of this map. This Creek was the water source used at the Flour Mill and Planing Mill shown above. The flour mill site was on the corner of Greenwood and South Second Street.

The farmers from the north end brought their grain to Enterprise in 1890. The following article explains. September 25,1890. "Lost Prairie Items - James Fleet returned from Enterprise recently with a load of flour for which he traded wheat of his own raising, receiving 35lbs. to the bushel. This is the maximum amount given at the mills, which surely speaks well for Mr. Fleet's wheat."

In 1895 ads were included in the columns of news events. Little two or three line blurbs like the following: "October 4, 1895 - Enterprise Mercantile and Milling Company - Bring us your grain. The Best Flour in Oregon."

Wallowa Chieftain, August 6, 1897

"About 50 feet of the E.M. and M. Co's mill flume collapsed last Monday. It will probably be several days before it can be repaired. There is no danger of a flour famine however as Mr. Bater says he has about 2000 sacks on hand."

Wallowa Chieftain, June 14, 1906

This advertisement appeared: "A case where the best is the cheapest Enterprise White Loaf Flour at 30 cents per sack or $3.20 per barrel."

Wallowa Chieftain, November 7, 1907

Mill Business Growing

"Regardless of the fact that the Enterprise M.& M. Co.'s mill is working over time every day, the demand for Enterprise Flour has become so great that they are unable to fill their orders. The mill is running 18 hours per day . . .

"Since the company has remodeled their mill and installed a sifter system their products have been second to none in the state and the consumers are fast finding this to be the fact . . ."

Enterprise Record Chieftain, January 5, 1922

Ships Million Pounds From E.M.& M. Co. Mill

"In the last month the Enterprise Mercantile & Milling Company has shipped a million pounds of mill products to Portland. This shows that one live local industry is doing much in the way of converting Wallowa county raw products into goods used in outside markets.

"Shipments have been particularly heavy for the last week, when six carloads of mill products went out. Three cars were loaded with flour and three with mill feed, all billed for Portland. The mill equipped with modern machinery, had a daily capacity of 120 barrels of flour, and handles a great quantity of wheat each year."

Enterprise Record Chieftain, June 12, 1924

"The Enterprise flour mill was sold last week by the E.M.& M. Company to George S. Craig and the Wallowa National Bank. Mr. Craig & W.R. Holmes, who leased the bank's interest will operate it as the Enterprise Flouring mill. V. I. Cronin will remain manager. The business will be pushed vigorously, buying wheat and producing flour and by-products and grinding all kinds of feed for customers, as in the past."

E. M. & M. Flour Mill Vic Cronin Senior worked here. Courtesy Mr. & Mrs. Vic Cronin

Enterprise Mercantile & Milling Co. Flour Mill Courtesy Enterprise Record Chieftain
The building down the street looks like it is under construction. Second Best occupies this spot today at 204 West Greenwood (1999). Notice the boardwalk way.

The map (on the next page) shows what the different rooms in the mill were used for.

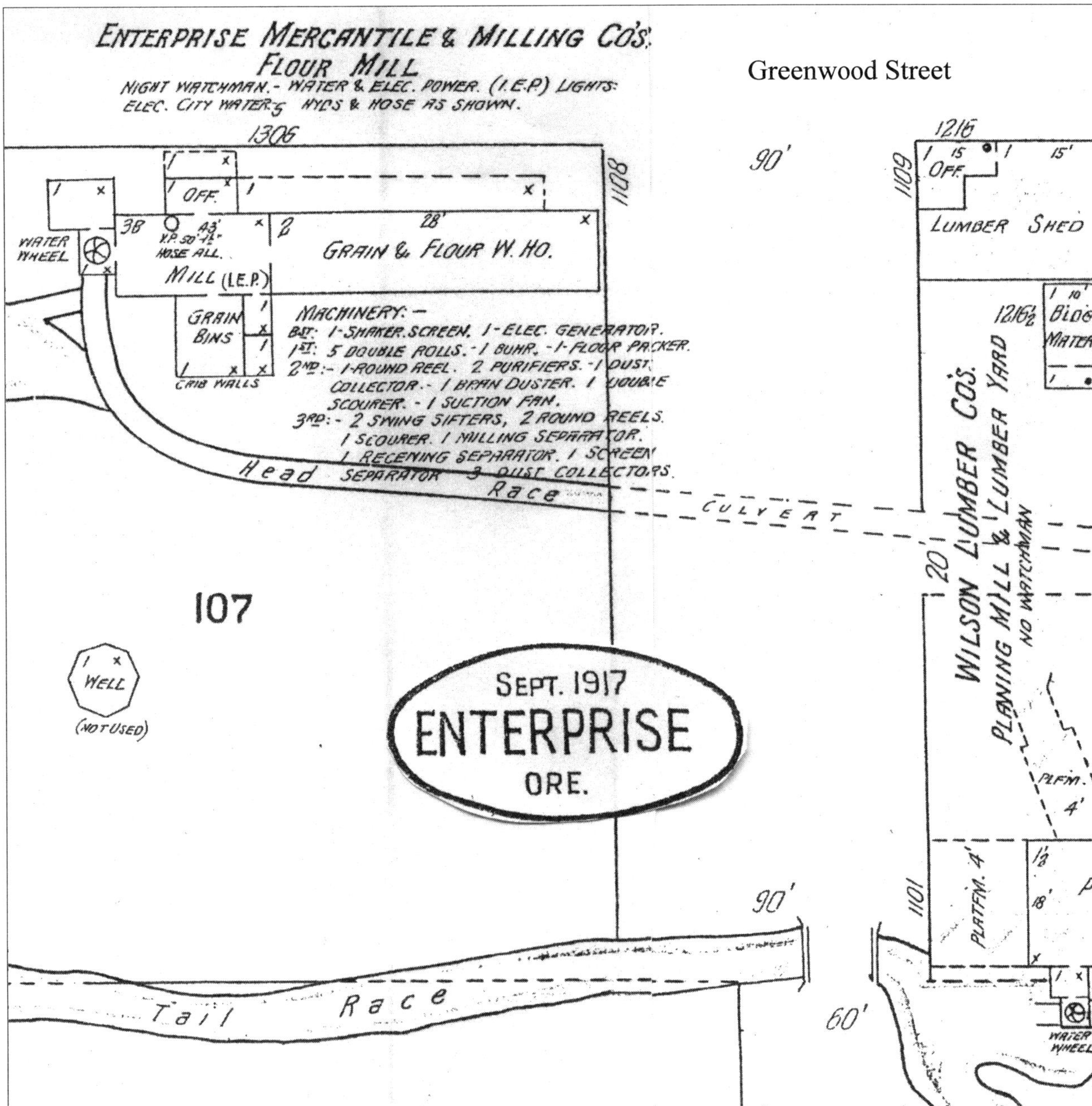

This 1917 Sanborn Map shows the flour mill had been enlarged after 1890. Most of the expansion was executed before 1907. Wilson's Lumber Co. on South 2nd St. included most of the block. The present business at that location today (1999) is the Enterprise Home Style Laundromat.

Enterprise Record Chieftain, August 21, 1924

"Enterprise Flour Mill Running after Overhaul Again, after a thoro [sic] overhauling of its water power plant. A new concrete foundation was placed under the turbine wheel and a larger pit provided for the discharge of the water. The flume was cleaned out also."

Enterprise Record Chieftain, December 4, 1924 (Advertisement):

"Any one doubting that we are making Hard Wheat Flour is invited to come to our mill and see the Hard Wheat going through the rolls . . . Don't pay $2.75 for a 50-pound sack of flour when you can get as good and better flour at $2.20. Save that 55 cents and keep the money in the county. Yours for fair dealing. Enterprise Flour Mill."

An advertisement in the same paper, March 5, 1925, stated:

"Three grades of flour.

"Calla Lilly, The regular Family Patent

"Hard - Marquis, The Bakers Patent

"White Loaf, The straight Grade

"Our flour is made from the best Wallowa county wheat sold under guarantee. We sell for less money than any outside flour of like quality. Enterprise Flour Mill. We believe in Reciprocity."

Enterprise Record Chieftain, December 23, 1926

"One of the first buildings on the site of the town of Enterprise was the flour mill erected by the I.C.M.& M. (Island City Mercantile and Milling Co.) Company in 1888 (at a cost of $14,000). Timber and lumber were sawed at the Meek mill on upper Prairie creek. For years the mill furnished flour in exchange for wheat and ground feed for farmers. Some dozen years ago the building was much enlarged and much new machinery added, making it the most efficient small flour mill in eastern Oregon. The capacity was increased from 40 barrels a day to 120, and flour was sold extensively in the general market. A couple of years ago the property was acquired by the Craig Family. The mill has been an important factor in the business of the community.

Enterprise Record Chieftain December 23, 1926

Flour Mill is Heap of Ashes

Fine Plant Destroyed by Night Fire to Be Rebuilt

"In the most destructive fire which has visited Enterprise for many years, the mill of the Enterprise Flour Mill Company was burned early Saturday morning with a property loss of about $45,000. The mill belonged to the Craig family and will be rebuilt.

"It is uncertain at this day whether the company now owning the property rebuilds, or another takes over, but there is no doubt that a new mill will arise next spring, as this is recognized as a community need. The property as it stands is valuable, with site, waterpower penstock and wheel, all ready for the construction of a mill. If the present company continues, some change among stockholders is possible, altho [sic] no decision has been reached on this point. Present stockholders, if they wish to sell, can get cash for their shares from other investors who would like an interest in the business.

"The new mill may not be as large as the old, but will have a capacity equal to the actual output of the old. So the plant will produce as much as the old and will give employment to as many men.

"When the fire was first seen a few minutes before 3 a.m., it was confined to the exterior of the north front of the building. The origin is naturally not known, but it is certain the fire began on the north, ground floor, and seemingly on the outside, Prentiss Homan and T.C. Gossett, who live nearest, agree that the flames were confined to that portion of the building when they were aroused and looked out.

"George S. Craig has a theory that the fire started in the pit under the driveway on the north, into which a cigaret [sic] might have been dropped by some boy coasting down the incline running west from the mill. This pit was well covered and dry and likely to be dusty, so that a smoldering fire could be hidden there all night.

"When the alarm was given, flames were leaping from the first floor walls for a distance of probably 50 feet. The response was immediate as volunteer firemen were scattered in their homes, and when the first stream of water was turned on the building, the fire was threatening the whole north wall and also was well started inside. The flames dropped when water was thrown on the mill, and spectators hoped the structure might be saved.

"However this was not possible, and after a few minutes of gloom during which clouds of smoke mounted into the night air, the red glow of flames shone thru the upper windows at the south. The building was on fire thruout [sic] the inside and could not be saved by any means at hand. Streams of water were kept playing on the mill at one place and another, but the building shortly became a great bonfire with flames shooting 150 feet into the air, lighting the town and valley and throwing out intense heat over a wide radius. In an hour, only a skeleton of red-hot burning timber stood upright and shortly they began to tumble into the heap within the area of the building.

"The warehouse at the east end received the attention of the firemen and was saved, altho [sic] the flour was damaged by smoke and water. Flames occasionally burst from the walls at the roof, and smoke pouring from beneath the floor showed much fire still there, well into Saturday morning. The portion saved at the first was not out of danger for many hours.

"The flour mill was built by the I.C.M.& M. Co. in 1888, the year the town was started. In those days a mill to grind the farmers' wheat and furnish stock feed was necessary for every new community and always was one of the first buildings erected. The lumber and timber for this building came from the Meek mill on Upper Prairie creek and pioneer carpenters and millwrights put the building together in a most substantial manner.

"For years it served its purpose well with minor improvements, and then, a dozen years ago, F.W. Hyatt completely remodeled structure and mill. The building was enlarged and the machinery was overhauled and more added to it, raising its capacity from 40 barrels a day to 120. During the world war the mill ground flour for export as well as for domestic consumption and did its part in feeding the allies. It was the most modern and most efficient mill in this corner of the state and for months ran day and night.

"Mr. Craig acquired the property about two years ago and other members of his family took an interest in it with him, including H.C. Mahaffey, manager of the business, W.F. Craig, and R.C. Wiley. By careful and efficient management the mill was made profitable and ran steadily, taking wheat in exchange for flour, and doing custom grinding, as well as manufacturing flour for the market. Its products were recognized as unexcelled.

"Early in the season farmers had several thousand bushels of wheat on storage in the warehouse, but all these holdings have been disposed of. Mr. Craig learned after the fire, and he breathed a sigh of relief when he found no patron would suffer any loss. Mrs. L.P. McCubbin, however, had 1000 pounds of seed oats which had been cleaned and left in the warehouse.

There was a stock of 800 barrels of flour in the east end of the warehouse where it got the benefit of the smoke and water. It will be good only for stock food.

". . . Mr. Craig took his loss in his usual philosophical manner, saying he hated to be out of business worse than anything else, and hoped to have a new mill as soon as a building could be erected and equipped."

The *Enterprise Record Chieftain*, December 23,1926 had an advertisement in the paper, "For damaged by fire flour sale at very low feed price." Enterprise Feed and Fuel Company, Wm. Snow and H. P. Rowe had bought the damaged flour for resale.

Courtesy *Enterprise Record Chieftain* December 23, 1926

Enterprise Mercantile & Milling Co. Flour Mill

"The E. M. & M. Flour Mill was erected by the I. C. M. & M. Co. . . Craig bought the mill in 1924. It burned in December 1926. This picture was probably taken after the remodeling was done in 1907.

The Wallowa Sun, December 23, 1926

FIRE AT ENTERPRISE RAZES FLOUR MILL

$35,000 LOSS ESTIMATED IN SATURDAY NIGHT BLAZE

"Built on the mill and covering a set of scales set in the driveway is a small shed and it was in this shed, entirely outside of the mill proper that the fire was first discovered late Friday night. When first noticed by a family living nearby, who turned in the alarm, the fire was confined to this small shed. In the few minutes required for the fire department to reach the scene the blaze had reached the main part of the mill and the entire west end of the structure was a mass of flames. . .

"Origin of the fire is doubtful although a carelessly thrown cigarette is believed to have been the cause. Leading from the scales down to the street is a slight incline and small boys had been using this as a coasting hill. It is thought that one of the boys threw away a cigarette in such a manner that it dropped through a crack in the floor and ignited chaff around the foundation of the scales." (Reatha McCormack, mentioned the incline at the mill was so steep she would always trot her horses down it when going in to pick-up flour or grain.)

The flour mill was never rebuilt in Enterprise after it burned.

WALLOWA ROLLER MILL

Lucian Cole was the original owner of the flour mill property. He sold (a portion) of his 160 acre homestead to George Post on March 8, 1889 (recorded in Book C Page 34, at the Wallowa County Courthouse). On March 9, 1889 George Post sold the land to the Island City Mercantile and Milling Company. This sale is recorded in Book D Page 33. Note that Post bought the land one-day and sold the very next day. It was five years before the Company built the mill. Post was evidently a land speculator.

The Island City Mercantile and Milling Company built the Wallowa mill in 1894. On November 26, 1896 the following transaction occurred: (Please see next page)

Records supplied by Frances Crow suggest that The Island City Mercantile and Milling Company sold it to The Wallowa Mercantile Co. on December 30, 1897.

Wallowa Chieftain, January 28, 1898

"Our flour and planing mills have been shut down for a few days on account of the ice in the ditch stopping the speed of the water and filling the valve wheels."

Know All Men by these Presents,

That the Island City Mercantile and Milling Company, a private Corporation, in consideration of its being the proprietor of the Town of Wallowa, in Wallowa County, Oregon, as laid out and platted by it in the NW¼ of NE¼ and the Northern part of the SW¼ of NE¼ of Section 14, Twp. 1, N. R. 42 E.W.M., in said County and State, hereby grants and dedicates the Streets and alleys therein to the use and benefit of the public according to the within platt and Map of said Town.

In Witness Whereof, the said grantor has hereunto signed its Corporate name by its President and Secretary, and caused its Corporate Seal to be offixed this 23rd day of November, 1896.

In presence of
E. P. Staples
John S. Clark.

Island City Mercantile & Milling Co.
By Chas. Goodnough, President.
A. R. Hunter, Secretary.

(Seal)

State of Oregon, } ss.
County of Union. }

This Certifies, that on this 23rd day of November 1896, personally came before me, a Notary Public in and for said County and State, Chas. Goodnough and Albert R. Hunter, President and Secretary, respectively, of the above named Island City Mercantile and Milling Co., and acknowledged to me that they executed the foregoing instrument as said officers of said Company for the uses therein named.

John S. Clark.
Notary Public for Oregon.

Wallowa Mercantile Co.'s Roller Mill

The Pacific Homestead, Salem, Oregon Thursday, October 19,1905

"The Wallowa Mercantile Company is the largest institution of a commercial character in the town. The company is composed of several of Wallowa's strongest business men and owns and operates a general store with a most complete stock of everything in their line, a modernly equipped flouring mill with capacity of over forty barrels. This company also owns and operates the electric light plant."

The following is more information from John Dierdorff's book on early electricity. "The first plant in the town of Wallowa was installed by the Wallowa Mercantile and Milling Company in 1903 and consisted of a 15-kw, 220-volt, direct current generator driven from the flour mill shaft. About the year 1907 a larger alternation current unit was installed with a larger water wheel, both of which were purchased secondhand from the company supplying La Grande and which had formerly been in use at the Orodell plant on the Grand Ronde river a mile or so west of La Grande. [See my book *From Trails to Rails* for more on Orodell.]

"This plant was bothered with ice troubles in the winter months and was often out of service for several weeks at a time. In June 1911 the Wallowa plant was purchased by George Jacobs."

George Evans was the first electrician at the Wallowa Flour Mill. "George Robert Evans, the third and last child of Alfred and Rebecca Jane Baker Evans was born in McKee, Jackson County, Kentucky 11 Oct 1860. He moved with the family to Arkansas in about 1870. His father died shortly thereafter and his mother married Cornelius G. Moore.

"George married Nannie Bowles in 1885 in Ozark, Arkansas. . . The morning after the marriage they left on horseback, Nannie riding sidesaddle, on a ten mile trip to visit Rebecca and Cornelius Moore . . .. George and Nannie rented a 160 acre farm in Madison County for a year before buying a farm that would be their home for the next five years.

"In March of 1891 George and Nannie were affected by "Oregon Fever" and moved from Arkansas to La Grande, Oregon where George worked in a brickyard for about a year. Nannie is quoted as saying "The wind blew all the time so the next year we came to Wallowa to get out of the wind and when we landed here, I said Thank God, I'll stay here." In Wallowa George and Billie Hamilton hauled freight on wagons from the rail head in Elgin to Wallowa. George, his brother James and Walter Alford bought a sawmill which was located on Mud Flat, later selling to Rod McCrae. Then George went to work running the E.A. Holmes flour mill for 14 years. Later, George worked for Nibley-Mimnaugh Mill until it was bought by Bowman-Hicks and continued with the new owners until his forced retirement after his 70th birthday. He had worked one of the hardest jobs in the mill, feeding the hog. The green stobs are picked up off the belt and put into a grinder where they are chipped up to be run into the boiler for steam fuel to run the mill.

"After living for a time on Bear Creek, George bought a lot in Wallowa and built a new home where the family moved on Thanksgiving Day, 1896. Their home was on a half city block where they also built a barn and had cows and pigs, a chicken house and a very large garden. They raised most of what they needed. With nine children to feed it was a challenge to provide for the family. George and Nannie were very frugal. The whole family worked at making the living.

"George and Nannie had ten children, five boys and five girls. James Alfred . . . Mary Frances . . . Effie Jane . . . Leonard William . . . Minnie Lulu . . . and a baby boy (born and died 27 Mar 1897), Georgia Emma . . . Homer R. . ., and twins Margaret and Morris (born 2 Nov 1905) . . .. Morris was killed in the Bowman Hicks mill 16 October 1925. George had very high principles and was very strict with his children . . ."George passed away at home 24 Jan 1942. Nannie passed away in a convalescent home in Emmett, Idaho 15 September 1953. They are both buried in Wallowa Cemetery." (Information by Nancy J. Henderson from information compiled by Elvyn T. Wolfe.)

Mary Frances married Elisha (E.H.) Wolfe. Their children were Evelyn, Elvyn, Frances, Wayne and Bill.

The following ads appeared in the *Wallowa Sun*. Evidently the Barkham's owned or operated the mill at this time. Then on November 12, 1914, the advertisement has the name of AF Dalrymple as owner or operator. I couldn't find any articles about the transfer of management or ownership to these people.

March 5, 1914

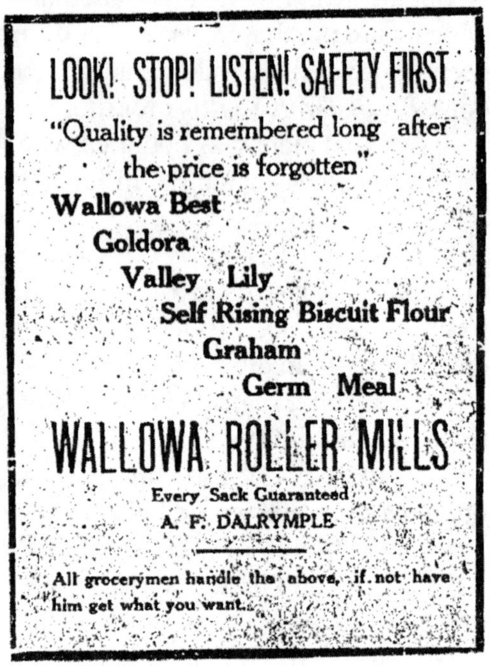

Courtesy of Wallowa Sun
November 12, 1914

J. O. KIDDLE BUYS WALLOWA FLOUR MILL

The Wallowa Sun, July 19, 1917

Deal Closed Tuesday and the Plant Will Grind Flour This Fall

"A deal which has been pending for a month or more was closed on Tuesday, whereby J. O. Kiddle takes over the 100 barrel flouring mill of the Wallowa Mercantile Company. The plant has been idle the past year, not being run since the lease of H.K. O'Brien and James Hayes Jr. expired about a year ago.

"Mr. Kiddle is a practical miller himself, having run flouring mills in Wallowa and Lostine for 16 years. He will have supervision of the plant and, besides grinding the flour, will do all kinds of chopping for stock feed.

"The mill will be overhauled and placed in first class condition before the 1917 crop is on the market for grinding. This mill is a big asset to the town and should receive the co-operation of all by liberal patronage."

Mr. & Mrs. Jesse O. Kiddle
Photo Courtesy Leland T. Kiddle

Mrs. Cora Tulley Kiddle was born in Wallowa in 1886. Jesse Kiddle came to Wallowa in 1902 to operate and manage the Wallowa Roller Mills. They were married in 1906. In 1908 they moved to Lostine where he operated the Lostine Milling Co. A year later he bought an interest in the mill and remained there for nine years. Returning to Wallowa in 1917 they owned and operated the Wallowa Roller mills for twenty-six years. His father had been a flour miller in England. In addition to operating flour mills, Mr. Kiddle operated one of the first electric plants in his area. Mr. Kiddle lived to be 94 and Mrs. Kiddle 106.

The Wallowa Sun, July 4, 1918

"The Wallowa flouring mill started grinding again Tuesday after being closed down for more than a week. The cause of the close down was on account of the recent high waters washing out a section of the Enterprise Electric Co.'s dam. The mill uses the water power when the electric company is not running their Wallowa plant."

The Wallowa Sun, December 26, 1918

MILL START UP

WALLOWA FLOUR MILL GRINDING AWAY

"The Wallowa Flouring mill, operated by Kiddle & Tulley, is grinding steadily away night and day and are tuning out on the average of 100 barrels of flour every twenty four hours.

"This means better than 400 bushels of wheat a day is being manufactured into the best flour on the market, which all of the local merchants handle.

"Since the ban on wheat has been raised, the demand has become greater, as most people were whole-wheat-bread hungry after dieting on substitutes for the past year.

"There are five employed at the mill, including the proprietors, Messrs. Kiddle & Tully, adding to the town's payroll, and making a good local market for the wheat raised in this section of the valley, and at the same time turning out a large quantity of mill feed, which finds a ready market from the dairying interests for winter feed for their dairy herds, which are increasing each year. This little mill is one of our big assets and deserves the patronage they are receiving from our merchants and fellow citizens."

BIRDS EYE VIEW OF WALLOWA, OREGON Courtesy Mary Lee Running

Picture was taken sometime between 1917, when first High School was built (visible as horizontal light rooflines at left margin- just behind the old Methodist Church and just left of the White School) and 1924, when first High School, which burned, was replaced by the one in use today.

Nibley- Mimnaugh built the Wallowa Saw Mill 1907-08. Bowman- Hicks bought them out in 1923. So I can't say for certain which was in business when the photo was taken.

The towering Bandstand was built in 1908 at the junction of Highway 82 and Whiskey Creek Road. Fire hoses and cart was stored under the bandstand.

Shown on the right is the Flour Mill. The inscription on the side stated The Wallowa Roller Mills Elevator.

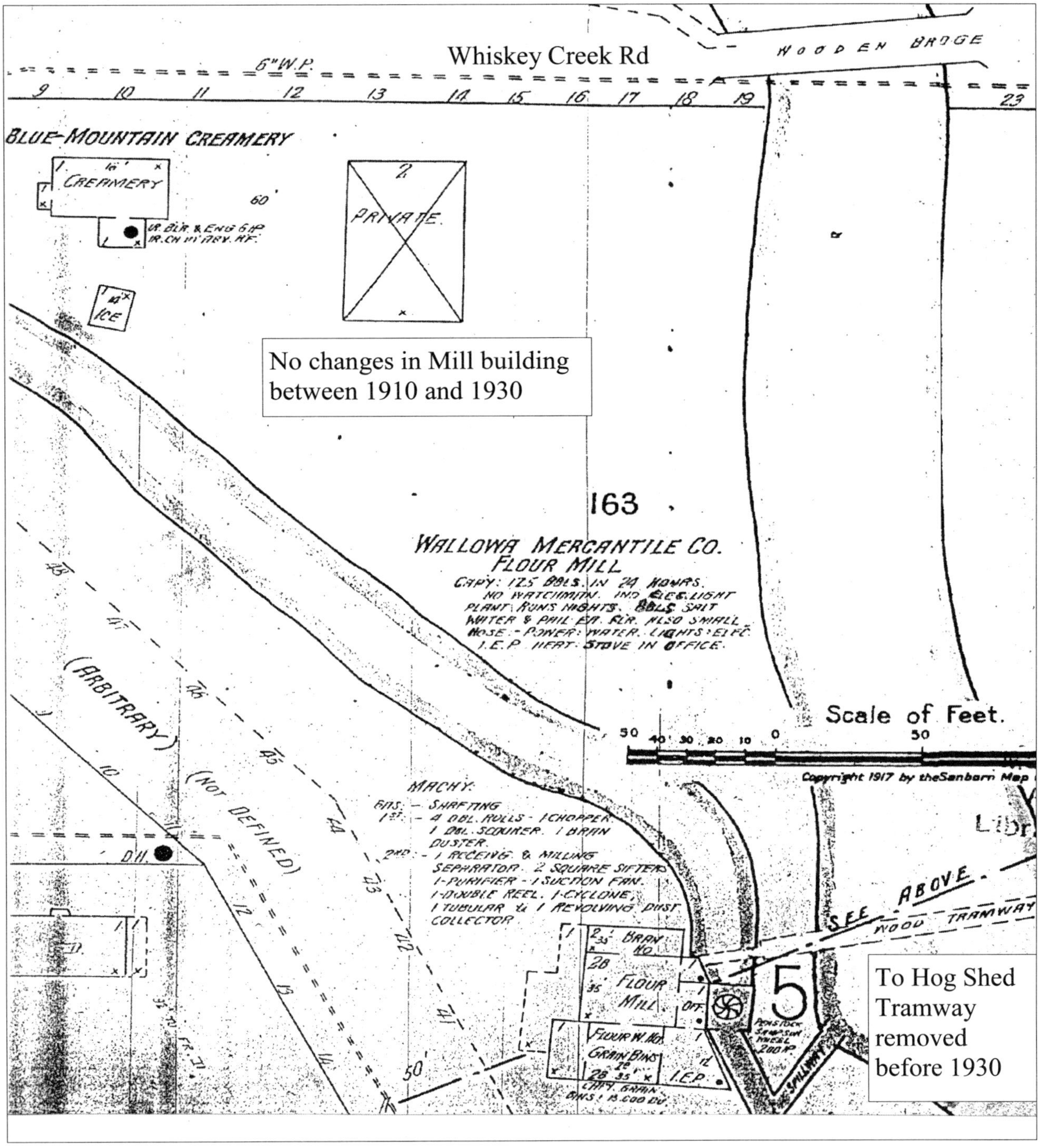
Whiskey Creek Rd
WOODEN BRDGE
6" W.P.
BLUE-MOUNTAIN CREAMERY
CREAMERY
ICE
PRIVATE
No changes in Mill building between 1910 and 1930
163
WALLOWA MERCANTILE CO.
FLOUR MILL
CAPY: 125 BBLS. IN 24 HOURS.
NO WATCHMAN. NO ELEC. LIGHT PLANT. RUNS NIGHTS. BBLS SALT WATER & PAIL EA. FLR. ALSO SMALL HOSE. - POWER: WATER. LIGHTS: ELEC. I.E.P. HEAT: STOVE IN OFFICE.
Scale of Feet.
Copyright 1917 by the Sanborn Map
(ARBITRARY)
(NOT DEFINED)
MACHY.
BMS. - SHAFTING
1ST. - 4 DBL. ROLLS - 1 CHOPPER. 1 DBL. SCOURER. 1 BRAN DUSTER.
2ND. - 1 RECEIVING & MILLING SEPARATOR. 2 SQUARE SIFTERS. 1-PURIFIER - 1 SUCTION FAN. 1-DOUBLE REEL. 1-CYCLONE. 1 TUBULAR & 1 REVOLVING DUST COLLECTOR.
BRAN HO.
FLOUR MILL
OFF.
FLOUR W. HO.
GRAIN BINS
I.E.P.
SEE ABOVE
WOOD TRAMWAY
SPILLWAY
5
To Hog Shed Tramway removed before 1930

WALLOWA FLOUR MILL c. 1910 Courtesy Winona Holloway

The following information is from the book *Moving On*, by Winona Holloway. "A lot of hogs were raised at our place and they were fed in good part by flour until the supply ran out. Yes, wheat flour! The Island City Milling and Mercantile Co. established business very early in the valley. In fact, it was not until a millrace was diverted from the Wallowa River, a mill set up and a store opened that the city of Wallowa amounted to much. During WWI the mill had a government contract and did very well. After the armistice the Wallowa Roller Mills, as the business was now called, was stuck with a ton or two of flour, which sat around for quite awhile; it collected moisture and caked up in the sacks. For some reason I have not been able to ascertain, it was called Red Dog Flour. Dad bought this for a song, soaked it and fed it to his hogs. The flour manufactured during the 20's and 30's was called White Rose. Many a sack was bleached and used for curtains, dishtowels and the proverbial female bloomers. I was never disgraced by having the name White Rose emblazoned on the seat of my pants but was humiliated when the elastic broke on my black-sateen bloomers and one leg stayed at ankle-length all day at school before I could rush home through town trying all the while to hold up my bloomer leg.

"During the Depression the flour mill was in bad shape financially. At that time quite a number of men pledged to bring in their wheat to keep it going. They were paid in cash or traded for flour. A certain percentage was charged for dockage for chaff and smutty wheat, as well as for the expense of milling. One percent or less was really good and five percent was not excessive. I have the ledger for the Wallowa Roller Mills in 1933-34. As I read the names I have visual images of these ranchers who brought in wheat from Flora, Paradise, Leap, Joseph, Enterprise and Imnaha . . .

"Jess Kiddle, the miller for many years, was a whiz at figuring in his head, and used to win mental arithmetic contests down at the pool hall. My brother Tom beat him a few times but no others he can recall. Hershel McGinnis was also a miller there and Bob Freels who was operator in the early 40's . . .."

OUCH! AND * # * + *

By Gene Hayes

"I was discharged from the U.S. Army the last day of the year 1945 and spent the rest of the winter trying to recover from malaria. Another problem, hanging over from World War II, was my right hand and left ankle was numb.

"In 1946 and 1947 I worked at several different jobs. The last week of August my friend and neighbor, Bob Freels asked if I would like to work with him in the Feed Mill, (former flour mill). The job was grinding grain for livestock feed. Sometimes I would haul grain from farms around the area. Sewing wheat and grain sacks closed took about three to five seconds to sew a sack.

"The State Industrial Accident Commission man from Salem was in the mill Friday, September 26, 1947 and Bob signed up for the insurance. Eleven A.M. the following Monday morning, 29th of September, Bob was out of town, a customer came in wanting some barley ground. I thought I knew the machine well enough to do the job. I commenced the rollers going. The rollers after a bit became too full and I made an adjustment, then reached to get a sample, going through a little door in the machine. (There was a metal plate I should have removed to get at the clogged up rollers). Sticking my fingers into the grain, I hit the rollers. My fingers still numb from my time in the war, I did not feel the rollers hit my fingers. My hand was crushed and hung up in the rollers. I kicked a lever and stopped the rollers. M. Henry, the customer, went to get some help. He ran to the Hunter Shell Service Station and got Roy Moores and "Henry" the old man that had made cereal there during the war years. The two of them figured out disassembling the roller. Raymond Southwick was summoned to take me to the Enterprise hospital in his car. The hospital then was in the former Wallowa County High School building.

"Dr. Ham transferred me to his car and headed for La Grande, old Grande Ronde Hospital, where that p.m. they did the amputation. I came out with a stub. At the old Enterprise hospital someone had been using the amputating saw, to cut plaster casts and wood. Dr. Ham didn't want to chance using that saw on a patient. He wasted no time on that drive to La Grande.

"I spent three weeks in the hospital with the hand amputation and was told I could go home. I was dressed and ready to catch the bus. Suddenly I got sick again, so "back to bed". My appendix struck and I was back in to get them removed. By the last of October I was out and going again.

"By April I had plans to go to the San Francisco Art School. I attended until August. Then went to O.T.I. now Oregon Institute of Technology. At O.T.I. I played football and the physical sport helped in my coordination. Changing from being right-handed to left-handed took patience and practice.

(Gene had enjoyed drawing since age four and often drew cartoons, etc., for family and friends while growing up.)

"It will be 50 years in September of this year, 1997, and the hand hasn't grown out an inch yet." (Gene, I congratulate you on the beautiful artwork you do today. See the article written about Gene in the June 5, 1997 *Wallowa County Chieftain*.)

Lela Locke remembers when Neal Knighton ran the mill in the 1940's and they made cereal. Bob Evans stated, "Merlin Waelty and James A. Evans also worked at the flour mill." The years they worked are unknown.

The Wallowa Record, December 22, 1955

FEED MILL HERE IS PURINA OUTLET

"Kirchgestener and Sons have been named local distributors of Ralston-Purina feeds for livestock and poultry. They received a carload shipment of the prepared feeds at their feed mill here December 12.

"Philip Kirchgestener moved here from Walla Walla the past fall with his family. He has been engaged in rehabilitating the feed mill machinery and building." . . .

The Wallowa Record, October 6, 1960

FLOUR MILL HAD BIG ROLE IN EARLY-DAY WALLOWA

"Wallowa's old flour mill, a commercial landmark since before the turn of the century is coming down. Only a few of today's Wallowa residents can recall the hey-day of this building's activities. One such is Jerry Maxwell, who sparked with reminiscence about the flour mill, when your reporter visited with him on his front porch Tuesday.

"It was built in 1894, and operated pretty steady till just after World War I," Mr. Maxwell opened his remarks. The operators were the Wallowa Milling and Mercantile Co., a local group of businessmen in a corporation, which did a business in those pioneer days as big as any in the county.

"They had the flour mill here, a large mercantile establishment standing on the corner where the Wallowa Cleaners now is, another mercantile establishment in Enterprise; as well as a hardware store and saddle and harness shop.

"In hard times when there was practically no money, we could turn in our wheat or our hogs and get everything we needed, almost, from the Wallowa Mercantile," Jerry recalled.

"They had a bridge across the river in those days, at the back of the flourmill. Across this they trundled the feed for a large bunch of hogs." They would buy the hogs, feed them out to about 250 pounds, and hire local teamsters to haul them to Elgin where they were put on railroad cars.

"Some idea of the size of the operation can be gleaned from Jerry's recalling a memorable trip hauling hogs to Elgin. That one winter day when Mr. Maxwell and his friend and fellow hog hauler, Bill McCrae, made the trip there were 100 four-horse teams and sleds on the road hauling hogs.

"There was two feet of snow on the road when they left Wallowa, and it took a good long hard day to get to Elgin with the 16 or 18 hogs in a load, weighing nearly two tons. The hog haulers stayed around Elgin the next days, helped load the hogs on railroad cars and started back home the third day. By the time they reached Minam all the snow had gone off the road. It took a good 12 hours to get the sled, by this time loaded with merchandise and implements for the Wallowa Mercantile, up the canyon.

"Going back to the hog feeding operation, the bridge was perhaps 80 or 90 feet long. The hand trucks used to carry the feed in sacks would haul about 400 pounds a trip. Later a set of light rails was put on the bridge and a four-wheel truck was used. The feed came in to the hog feeding shed on a second story level. There it could be dumped from the sacks into self-feeders for the hogs on the ground level.

"The flour mill would buy about 75 percent of the wheat raised in the area around Wallowa. "They made all the flour we could eat, and shipped a lot out," Jerry recalled. Often times the mill would be running night and day. It was not a large mill, and two men were enough to keep it running.

"The lumber from which the flour mill was made was milled in Union County, because at the time mills with enough capacity to fill an order of this size had not got to operating in the Wallowa area. "If there's a knot in any board down there, it is a board that was put there after the mill was originally built," Mr. Maxwell surmised.

"Jerry was a neighbor of the flour mill, in that he operated his packing plant just to the north about where Wayne Prince now lives. He also had his meat market where the post office now is, and his slaughterhouse on the place where Roy Carper has lived for so many years. The meat products brined and smoked at the packing house were sold to the markets of the county.

"Mr. Maxwell recalled that Gene Holmes, Charles Hunter and his father Bill Hunter were among the stockholders in the corporation which operated the flour mill. Charlie Goodnough and J. M. Church were also stockholders and also owned the mercantile and milling company at Enterprise. Ed Goodnough, a nephew of Charlie, was later the head of the company. He was a stickler for honesty and fair dealing.

"The flour mill was operated by water power. The town had no central supply system, but the flour mill put in a water power ram which pushed water from the mill stream into a reservoir sitting near the present Bill Maxwell home. Several homes in that area where served, and there were several hydrants installed along Main Street and serviced from this reservoir. So far as Jerry can recall, no one ever got a bill for the water service."

The home of Mr. And Mrs. Tim McDaniel at 503 E. Highway is at the location where the old flour mill stood for so many busy years.

The millrace was in front of the Lawton McDaniel home. One crossed a bridge to get to their house. Donna McDaniel (Skovlin) Mary Lee Maxwell (Running) and other children had fun swimming in the millrace.

LOSTINE FLOUR MILL

"The Lostine Flouring Mill Company was organized in 1895 by local business men and farmers and a fine, fifty barrel mill was erected. After one season's run, however, in December 1897, this important enterprise was destroyed by fire, entailing a serious loss upon the community. But, the need of a mill was so urgently felt that another company was immediately organized by local capitalists and a mill having a capacity of seventy-five barrels was erected in 1898, . . ." (Information is from *The Illustrated History of Union and Wallowa Counties, 1902*.)

Wallowa Chieftain, April 8, 1897

"Plenty of flour in the mill now and your grist can be exchanged while you are unloading. The mill is so full of flour that one can hardly find the miller. The roads have been so bad this winter that Mr. Rumble could not get his flour down to the railroad."

December 3, 1897

"The flouring mill is being run day and night and several loads of flour leave Lostine everyday for outside markets."

The following articles were written, in the local paper, anticipating the building of the (second) mill after the first one burned.

April 1, 1898

"All the stock has been subscribed and a new milling company will be incorporated in ten days."

April 29, 1898

"The flouring mill project needs stirring up. It has been quiet of late."

May 6, 1898

"I.T.D. Lacert was awarded the contract for furnishing lumber for the Flouring Mill and has begun delivery."

THE PACIFIC HOMESTEAD Salem, Oregon October 19, 1905

"The flour mill of Lostine is owned and operated by Mr. C. Smith. It has a capacity of eighty barrels per day and runs continuously in order to supply the local demands. Mr. Smith is also owner of the Lostine electric light plant which furnishes the city with lighting capacity."

When checking records, several transfers of the Lostine Mill were found: U.S. Patent recorded to Robert L. Erwin for the land where the mill was situated. Then the records showed the following transfers: to S.L. and Florence Magill then to The Lostine Milling Co. July 21, 1898. There is a reference to R.B. and H.W. Cole to Lostine Flouring Mill Co. Another transfer recorded to John H. McCubbin President and G.J. Bowman Secretary acting for Lostine Milling Company. Later a transfer was made from Lostine Flour Milling Company to Barnard & Leas to satisfy the Mortgage. That board of Directors sold to W.W. White.

Courtesy Ethel Crow

LOSTINE FLOUR MILL Postal Card Date October 1, 1914 from S. E. Hambleton to Gussie Crow

Wallowa Chieftain, June 14, 1906

An advertisement in the paper stated: "The best, Lostine Flour now selling at $3.50 per barrel or 90 cents per sack."

Wallowa Chieftain, June 27, 1907

"A deal was closed Monday of this week whereby E.E. Funk, Sam Litch & R.W. Fitzpatrick became the proprietors of the Lostine Flouring Mill. The mill was formerly owned by C.M. Smith & F.D. McCully. The new owners will continue to operate the plant in the name of the Lostine Milling Company.

In addition to the flouring mill, they also bought the Lostine Electric light plant and the C. M. Smith residence property. The new firm will assume charge of the business about August 1st."

Wallowa Chieftain, August 15, 1907

"The Lostine Milling Co. has secured the service of Mr. J.O. Kiddle, who for several years has had charge of the flouring mill at Wallowa, to run the mill at Lostine. They are fortunate in securing Mr. Kiddle as he is an experienced mill man and is well acquainted with the people of the county as well as being thoroughly familiar with the conditions of the country, having had full charge of the Wallowa mill for the past five years and will need no introduction to the people of Lostine or to the patrons of the mill from otherparts [sic] of the country. We are informed by Mr. Litch, one of the firm that they will start the mill to running just as soon as threshing begins and wheat can be delivered at the mill . . . They will endeavor to increase the output of mill products and will even look to it that Lostine flour remains where it was placed by the **Lewis & Clark Exposition** judges - in the gold medal class."

J.O. Kiddle is called a miller at Lostine, in an article dated May 6, 1910. The article listed the merchants that were staying in Lostine and not moving to Evans. Several businesses moved from Lostine to the town of Evans when it was formed in 1910.

Town of Lostine

Courtesy Rhee Caudle Lathrop and Arlene Caudle Makin

This panoramic view shows the flour mill on the far right. The Lostine Presbyterian Church is close to the left side. This picture was probably taken soon after the mill was built in 1895.

NOTICE

ON and after September 1st the Lostine MILLING CO. will conduct its business on a strictly cash basis. All back accounts will be due at said time and must be paid in cash, hogs or grain not later than October 1, 1912. Thanking our customers for past favors and soliciting your future patronage to the same extent on our cash basis, which after experience and careful consideration we have decided is better for all parties concerned.

We remain very respectfully,

Lostine Milling Co.

INCORPORATED

There is a possibility that the mill closed down after Mr. Kiddle left in 1917.

The mill was turned into a Land Plaster Mill in 1926 as the following story explains.

MARBLE COMPANY BUYS OLD FLOUR MILL

Enterprise Record Chieftain, March 25, 1926

Lostine Plant will be used for grinding Land Plaster

"Another forward step was taken last week by the Black Marble and Lime Company in the purchase of the old flour mill at Lostine, which will be converted into a plant for the production of land plaster. The mill was brought [sic] from William Kennedy of Portland, who acquired it from the Wallowa Milling and Grain Company when the latter was liquidating its assets. Judge L.E. Jordan, H.J. Kressly and A.J. Warren went to Portland to make the purchase.

"The mill was completely equipped for making flour. Part of the machinery, suitable only to grinding and preparing flour will be sold, and part will be retained for use in producing land plaster."

Years ago Link Magill told me that someone raised mushrooms in the lower part of the flour mill. I remember going to the lower level of the mill to have a look. There were boxes 24" wide x 30" deep and 8" high, all in long rows. Jerry Moffit at Moffit Brothers Tire Service in Lostine, Oregon uses these boxes for storage of hubcaps, etc. I wonder if anyone remembers anything about this mushroom farm?

This would have been the third mill, a much smaller mill than the first two. I can't find a picture of this mill. Mr. & Mrs. Jerry Moffit tore down the old mill, and today (1997) their home is at that spot.

FLORA FLOURING MILL

The News Record, Enterprise, June 25, 1908

"The North Country to have a flour mill. A Company incorporated with $25,000 Capital Stock to erect a flour mill at Flora, to be named Flora Flouring Mill, Inc.

"It is to be a two-story structure with a basement, located on the north west corner of the James Bingaman ranch. Work has started. It is to have a capacity of 50 barrels a day. The machinery was estimated to cost $6,000.

"Directors and officers: President, J.Q. Connally; Vice President, W.H. Baker; Secretary, Lincoln Austin; Treasurer, Paul Moore; Managers, F.S. Johnson, R.M. Martin."

From the *Flora Journal*, June 1908 - "Excavation is being made for the basement of the mill building and rock is being hauled for the foundation. R.W. Barkham arrived from Weston, on Wednesday stage. He will oversee the putting up of the building."

July 1908 - "The material for the flouring mill is partly on the grounds waiting the planer. The engine and mill are on the grounds. An expert miller, Mr. Markham, of Nebraska will supervise erection and installation of machinery. There is a fine big spring to supply water for the boilers."

July 28, 1908 - "Wagon Freighters W.J. Akin, Dode Conrad, O. Gowey, have arrived at the Sled Springs station with the engine and boiler for the Flouring mill. The planer has arrived and is to be ready to plane in two weeks."

Aug. 20, 1908 - "The work is progressing at the mill. The planer is nearing completion."

Sept. 20, 1908 - "THE FLORA FLOURING MILLING COMPANY, INC. Business title has been filed in Salem, Directors and Officers: J.Q. Connally, W.H. Baker, T.M. Gilmore, Paul Moore, F.S. Johnson and R.W. Markham."

Nov. 14, 1908 - "R.D. Meeker and Nels Hansen, millwrights have placed the machinery in the flouring Mill. Mr. Hansen and family are recently from Milan, Washington."

Feb. 19, 1908 - "The mill is grinding wheat and making flour. The flour is sacked with the label 'PINE BUR'."

"While the flouring mill started out as a corporation, it eventually failed and Mr. W.H. Baker was sole proprietor. Mr. Baker also owned the sawmill.

"The Flora Flouring Mill was a definite improvement for the North End farmers. They could exchange their wheat for flour for human consumption and ground wheat for stock feed. The business functioned for about 25 years; however, during the last years, its grinding was sporadic, principally after harvest in the fall. The trend in the early 1930s was to truck the bumper wheat crops to the railroad in Enterprise.

"Nels J. Hansen 1868-1941 - came from Denmark in 1868 when he was 18 years of age. He learned English, educated himself, and became a cabinet maker and millwright. He built mills and elevators in Missouri and moved to Spokane, Washington with his family. When he learned Flora, Oregon needed a millwright, they again moved.

"In 1909, they settled in Flora, where Nels built the mill, managed and operated it for about 14 years. He became postmaster in 1923. He operated a grocery store in connection with the post office until 1937, when he sold out and moved."

The above information is from Lorlene Beddow's book *Carving the North End Wilderness:* Flora" 1985.

Courtesy Beulah Wynans

Above is the **Flora Flour Mill** building. It was in front of the Bill Baker sawmill.

Courtesy Joyce Clark McFetridge

Community of Flora

This picture was probably taken soon after the mill was built in 1908. The flour mill is the large building near the big tree. Mr. Charlie Bell was a mail carrier and kept his horses in the barn to the left of the mill. George and Nellie Clark built the house on the left of the barn. It is known as the Conley house. The Methodist church is between, (in back) the Clark place and the mill. Mr. Bill

Baker owned the large store behind the big tree. Located almost in the center was the Austin Hotel. It had a porch. The large building, on the right, with the dark roof is the Christian church. Mr. Frank Johnson platted the town of Flora in 1897. His house is on the far right.

Information about the location of the mill came from Joyce Clark McFetridge. "It was on Buzzard-Lane. The sawmill and pond were east of the flour mill. Both plants used the same water. In the winter the pond would freeze and the ice was cut or sawed into blocks and stored for summer use."

Probably the last year the mill operated was in 1932, according to Merritt Holloway. There was still flour there in 1933. An incident Merritt remembers is about a field trip when Mr. Easton, the superintendent of Flora High School, took the student body to see the running of the mill. Daring students went hand-over-hand on the two-by-six ceiling joists that supported the overhang over the driveway to keep the area free from snow. Brady Botts tried it and fell to the floor breaking both arms. Leonard Murrill remembers, "that he and friends played in the mill. It was in the thirties that the mill stopped making flour. Mr. Pittner tore the mill down sometime after World War II. Joe Shuman used the foundation of the old flour mill when he built his new house".

TROY ROLLER MILLING CO.

Town of Troy Courtesy Joe Beach

Post card dated Nov. 15, 1913. The store and Post Office are on the left side. In the trees behind the store is the schoolhouse. The impressive flour mill building is on the right side.

"The flour mill at Troy started operations in the fall of 1912. It was built by the men of the district who could see no reason for hauling their grain out and their flour in. On the benches above the Grande Ronde and Little Salmon rivers, much wheat is raised, along with other grain and fruit and a great variety of vegetables. But the canyons are so deep that freighting is a big problem, and whenever the haul can be saved, there is a great gain. The mill is run by water from the Little Salmon River. The photograph reproduced here was taken on the bank of the stream across from the mill.

"Before the steel bridge was built across the Grande Ronde River at Troy a ferry was in service there. The ferry is at the bottom of the famous Troy hill where a road drops from the upper levels of Lost Prairie into the depths of the canyon."

"Lon (Alonzo) Silver owned the mill. It was a good business, until after World War I. During the war the government had all the flour tied up, as they were shipping it to Europe to feed the troops. After the war was over all the flour mills were left holding the bags. The bags full of flour that is. The flour spoiled in the mills. Mr. Silver went to work up on Grouse Flat for the farmers. He needed money to make payments on the mill."

"Arlie Cook Hopkins painted the mill white. Large black letters on the front of the mill stated, TROY ROLLER MILLING Co.

"Weldon Silver bought the mill from Lon. Weldon sold the mill to Les Harding in 1922.

"People from town brought buckets to the mill water wheel and filled them with water for their gardens." The above quotes are memories of Ralph Silver, the son of Weldon.

Enterprise Record Chieftain, June 30, 1927

"Part of the dam at the Troy flour mill was washed out early in the spring and will be rebuilt shortly. The dam is across the Little Salmon (Wenaha) river from, which is taken water to turn the wheels of the mill. L. J. Hading is proprietor and manager of the mill."

November 30, 1939

"The last of the 3-story flour mill was taken down during the week and part of the lumber hauled to the Jim Fordice place at Lost Prairie."

TURN BACK FLOUR AT NATION'S CALL

Wallowa County people answered the call when the government asked for the citizens to turn back flour, to be sent to the troops overseas during World War I. I found several lists, in the paper, of the people who turned back flour. It was a national effort.

Enterprise Record Chieftain, June 6, 1918

People of County Start Movement to Help Feed Soldiers and Allies Overseas.

"In response to the appeal of the government for the turning back of wheat flour, the movement is getting under way in Wallowa County on a scale which really counts. Nearly every store has started a list of returned flour. Far the greatest bulk is received from farmers who had stocked up last fall, after the long established custom of delivering wheat at a mill and receiving back part of it in flour."

"The largest single turn back in the county is credited to Ernest F. Johnson of Wallowa who handed over 28 barrels. For some years Mr. Johnson has used 35 barrels of flour a year in his whole establishment, and last fall he got that amount. Then came the campaign to reduce the consumption of wheat products and to use more substitutes, and Mr. Johnson followed instructions so faithfully that only seven barrels of white flour were consumed, and now he has turned back the balance.

"Charles T. White of Imnaha responded promptly to the call and told A.C. Miller, county food administrator, he had 6 sacks or 300 pounds which he would turn over to the store at the bridge.

"S.A. Blevans brought back 8 sacks from his farm on the Divide, and S. P. Williams turned over five sacks from his place on Upper Prairie creek. Announcement was made last week of the pioneer in this movement, H. Wilder of Flora who, by dint of saving had five sacks on hand which he wanted sent direct to Picardy.

"Enterprise stores yesterday reported the following to have turned back wheat flour: C.D. Emmons, E.G. Cummings, W.H. Weaver, W.F. Savage, G.P. Cheney, S.P. Miller, A.E. Hartley, Frank Walker, Al Thompson, W.T. Bell, J. S. Kay, W.F. Craig, H.A. Quesenberry, C. F. McPherson, Dr. J.H. Thompson, R.H. Danley, W.E. Warden, Mrs. H. Rodgers, Daniel Boyd, Mrs. R.R. Thompson, Geo. H. Holmes and S.L. Burnaugh.

"The grocers are anxious for the government to take over their stocks of white flour, so as to close up the matter and stop inquiries. Mr. Miller expects this will be done in a very few days. The merchants and warehousemen will ship to a government agency and make a sight draft for the value of the flour.

"One store, the E.M.& M. company, has removed all wheat flour from its retail establishment, so that no one can be reminded of them. The flour has been taken to the company's mill, where the turn-back shipment is being assembled.

Bring Flour to Joseph

"Since the call was made last week by A.C. Miller, county food administrator, for the supplies of flour in excess of one sack or over, the people of Joseph and the farmers of the surrounding territory have responded. In all over 33 sacks have been turned in at the stores and the flouring mill."

Enterprise Record Chieftain, June 13, 1918

FLOUR WILL START FOR PICARDY SOON MERCHANTS STOCK & SUPPLY TURNED BACK BY CONSUMERS TO BE TAKEN OVER BY U.S.

"Wheat flour in stores and warehouse, and that returned by consumers, will be taken over by the government by June 20, according to word received by A.C. Miller county food administrator. It will be shipped to a Pacific port, whence it will go to its destination in one of the new ships. Merchants turning back flour will draw on the government for its value at the time of shipment.

"The flour taken back does not include graham and whole wheat. These are not suitable for shipment to Europe. "The flour saved by H. Wilder of Flora was brought to Enterprise yesterday by C.F. Evans and turned over to the Woolgrowers Warehouse for shipment to Picardy (England). As was Mr. Wilders wish. Other returns of flour during the week were made by Charles A. McAlister, George Emmons, W. D. Forester, J. A. Whitman, A.W. Haggerty, G.W. Hyatt, C.D. Emmons, D. W. Tierney, J.S. Surber, H.C. Hearing, John Vonderpoel, Charles Whitehead, Guy E. Skaggs and Ralph Williams. "Surplus flour still coming into the stores at Joseph, H. S. Hunt, William Scott, A. E. Gibbons and A. K. Parker."

Wallowa Sun, July 4, 1918

Use Substitutes and Help Conserve Wheat
Good Combinations Given

"Beginning with Friday, July 5, Oregon lifts the ban on wheat and returns to the 50-50 Wheat and substitutes. In order to facilitate matters the U.S. Food Administration have planned a card giving the equivalent in weigth [sic] of one cup of wheat as follows:

"Barley, 1 3/8 cups; buckwheat, 7/8 cup; corn flour, 1 scant cup; coarse corn meal, 7/8 cup; fine corn meal, l scant cup; cornstarch, 3/4 cup; peanut flour, l scant cup; potato flour, 3/4 cup; rice flour, 7/8 cup; rolled oats, 1 1/2 cups.

"Using the above measurements one can use any recipe that they are accutsomed [sic] to and put in the right amount of substitute.

"Better results can be obtained with the mixture of two substitutes than with the straight use of one alone but not every two can be combined with good results. Some good combinations are:

Rolled oats (ground)	Corn flour	Barley flour	Rice flour
Peanut flour	Potato flour	Soy-bean flour	Corn meal

"Any of the substitutes in the first list can be combined with any in the second but do not combine only those in the one list.

"With the substitutes one can not always tell about the looks of the batter. It is better to use accurate measurements. All substitutes should be baked slowly and longer. Drop biscuits of substitutes are better than rolled.

"A copy of this card for hanging in the kitchen for ready reference can be had by addressing the U.S. Food Administration, Washington D.C."

I found the above article interesting but I am certainly glad there is plenty of wheat flour today. Probably the cooks in 1918 were glad to have the substitute information.

These men are loading sacks of flour into the wagon. This picture was enlarged and taken from the Lostine Post Card dated 1914 from Page 36.

FLOUR SACKS

These flour sacks had white backgrounds. The stripes were blue and red. Yellow and rust flowers with green leaves adorned the next sample. Last is a great Hawaiian design with red-yellow hibiscus.

The industrious housewife found many uses for the cotton sacks in which flour was sold. Everyone who lived in the era of cloth sacks remembers all the articles that were made from them: dish towels aprons, pillow cases, quilts and even sheets. Many a mother and little girl had their undergarments made from the adorned sacks. The sacks had a variety of vivid designs that advertised the products contained within. Often sacks were boiled, with soap and lye added, to eliminate the design and printing.

Courtesy of Ruth Dressler

FLOUR SACK FROM WALLOWA ROLLER MILL
WALLOWA, OREGON

49 Lbs.
When Packed

Bernice McBath gave me the flour sack with this design.

FLOUR SACK FROM WALLOWA ROLLER MILL
WALLOWA, OREGON

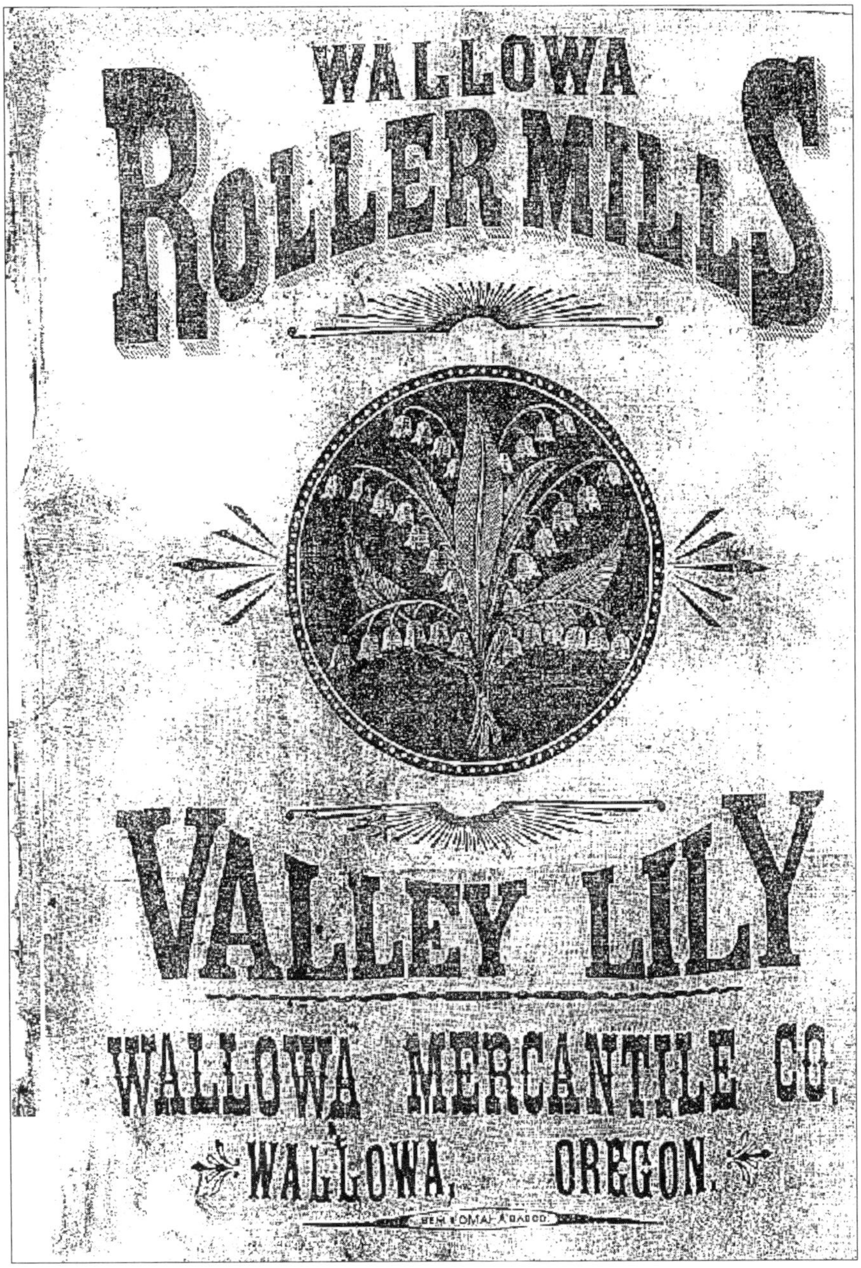

Courtesy Bobbie Long

FLOUR SACK FROM WALLOWA ROLLER MILL
WALLOWA, OREGON

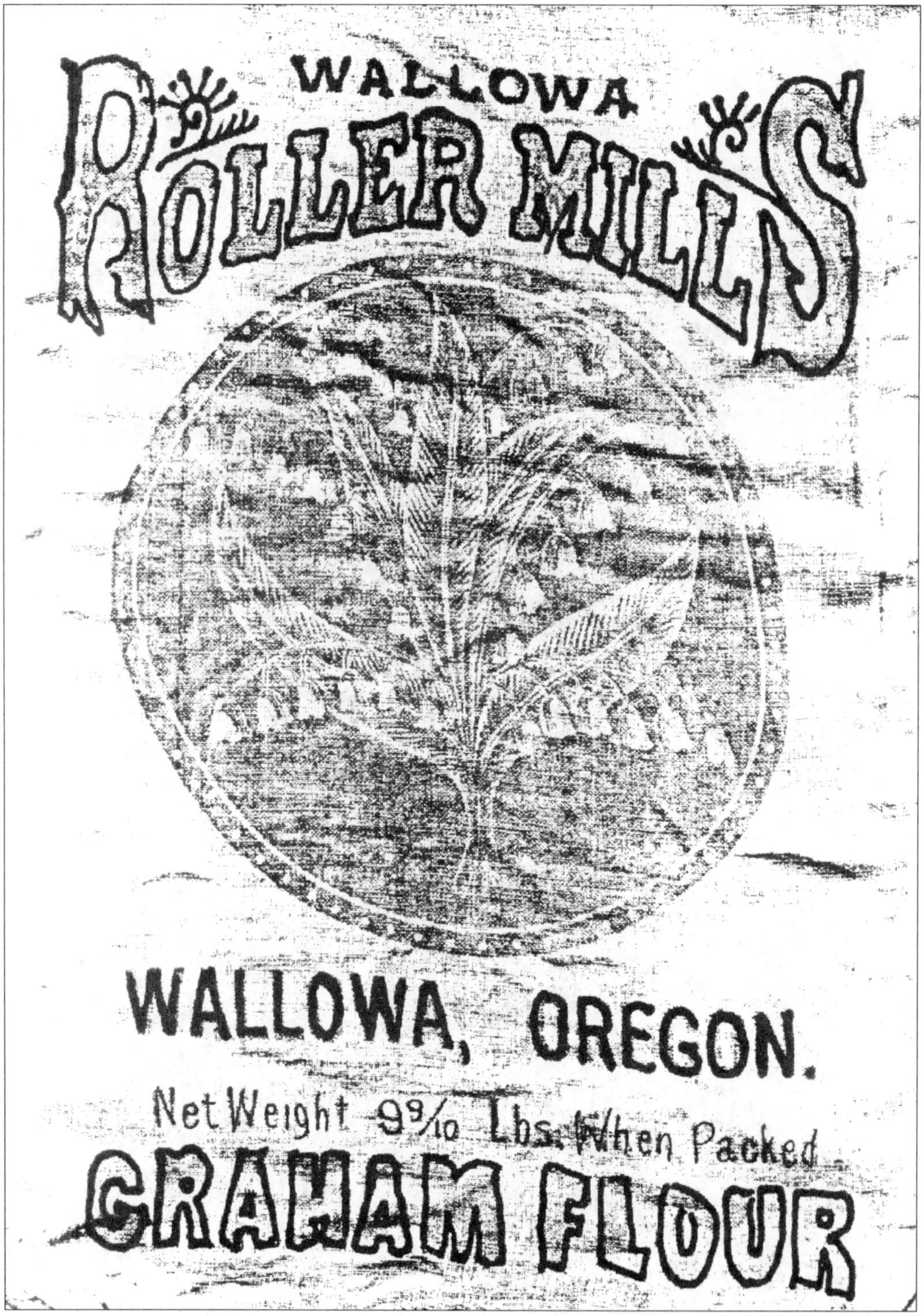

Courtesy of Jo Ann Snead

FLOUR SACK FROM JOSEPH MILLING COMPANY
JOSEPH, OREGON

Courtesy Moonlight Graphics Co.

FLOUR SACK FROM FLORA MILLING COMPANY, FLORA, OREGON

The flowers and the lettering Mountain Lily Flour were printed with red ink. This type of sack was probably used from 1908 until the mill closed in the 1930's.

Flour sack sample above was off-white background with brown dots. Squares are aqua and gold plaid.

Courtesy Beulah Wynans

FLOUR SACK FROM WALLOWA ROLLER MILL
WALLOWA, OREGON

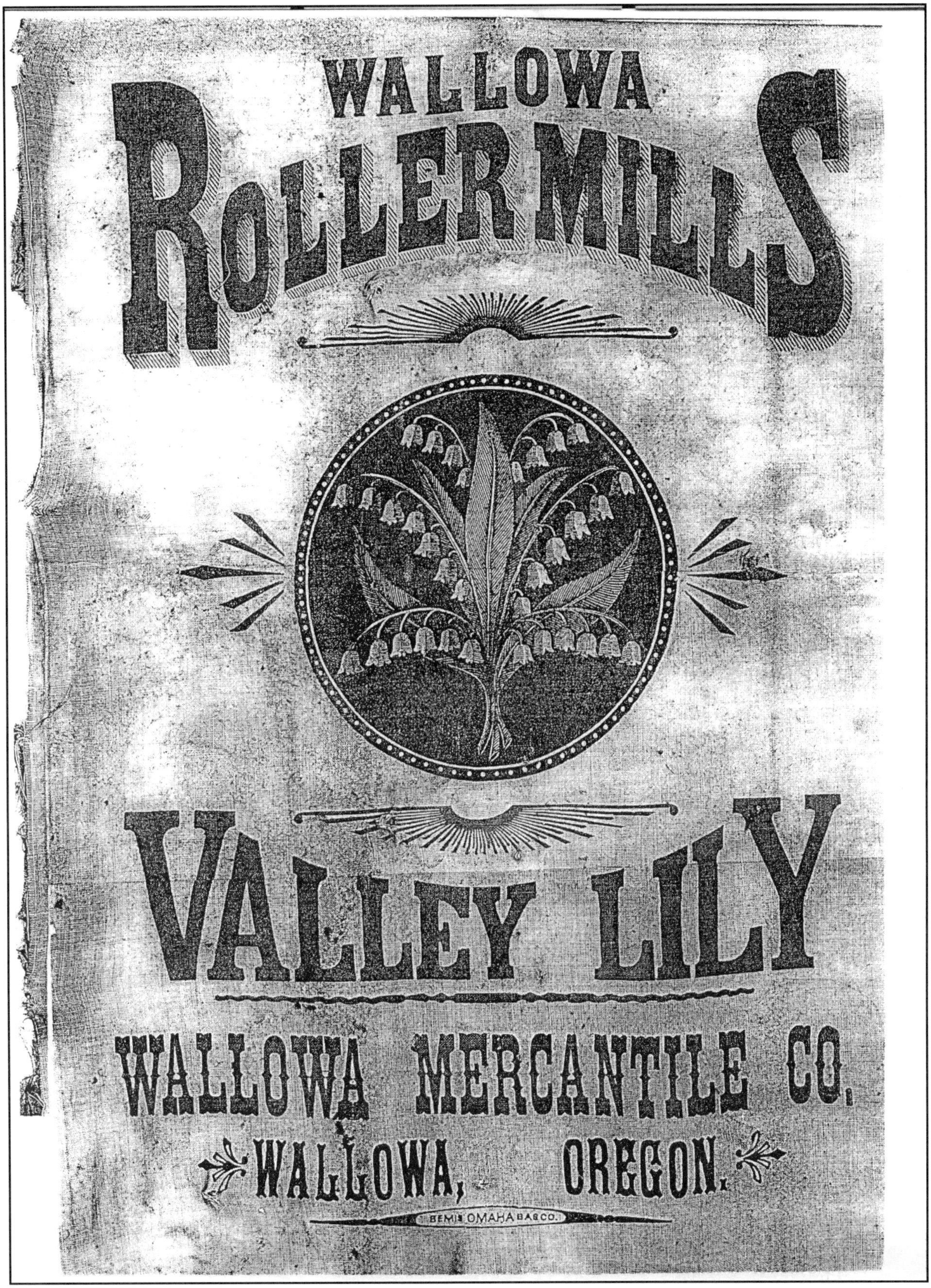

Courtesy Jo Ann Snead

FLOUR SACK FROM ISLAND CITY, OREGON
UNION COUNTY

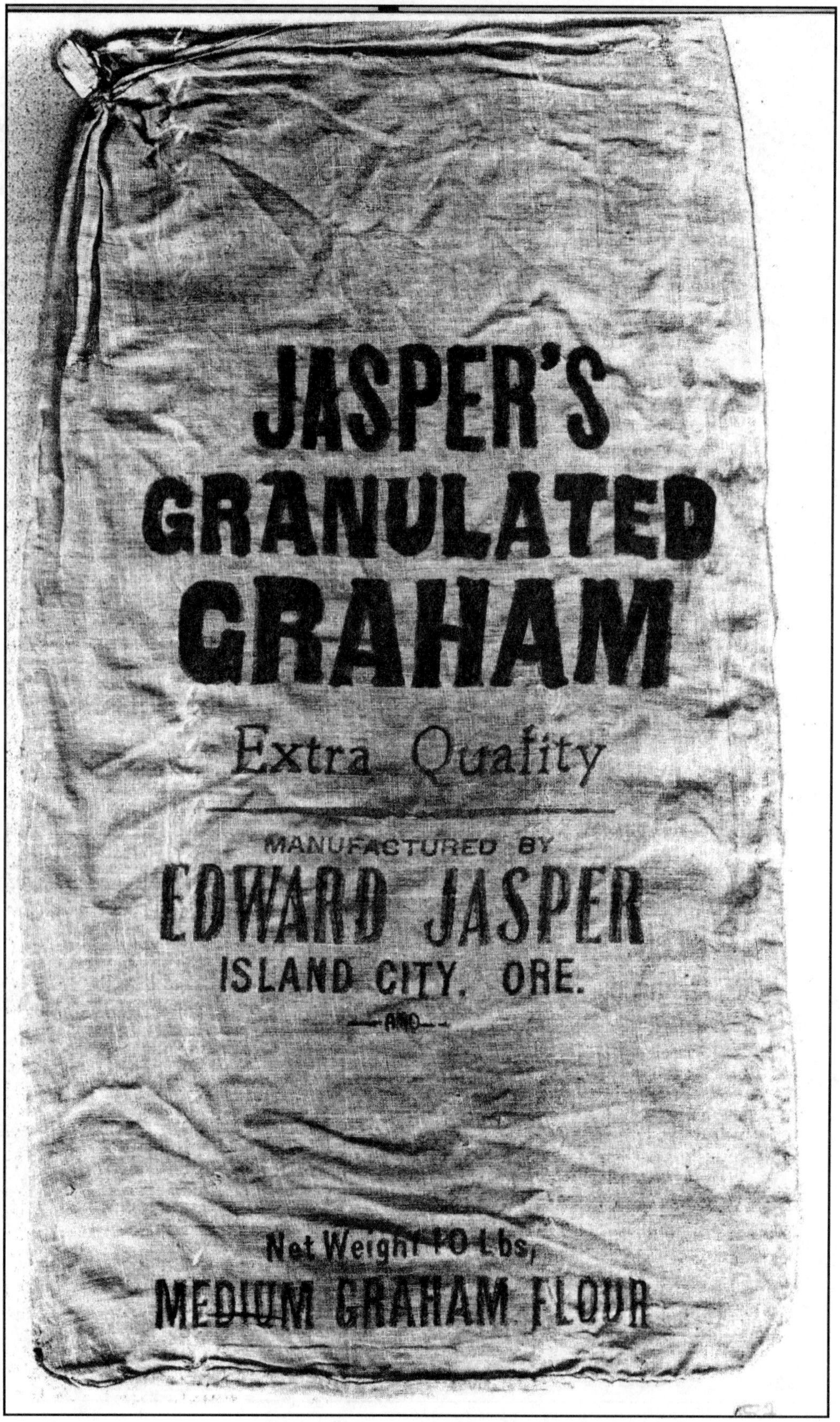

Courtesy Jo Ann Snead

FLOUR SACK FROM ISLAND CITY, OREGON
UNION COUNTY

Courtesy Jo Ann Snead

FLOUR SACK FROM PENDLETON, OREGON
UMATILLA COUNTY

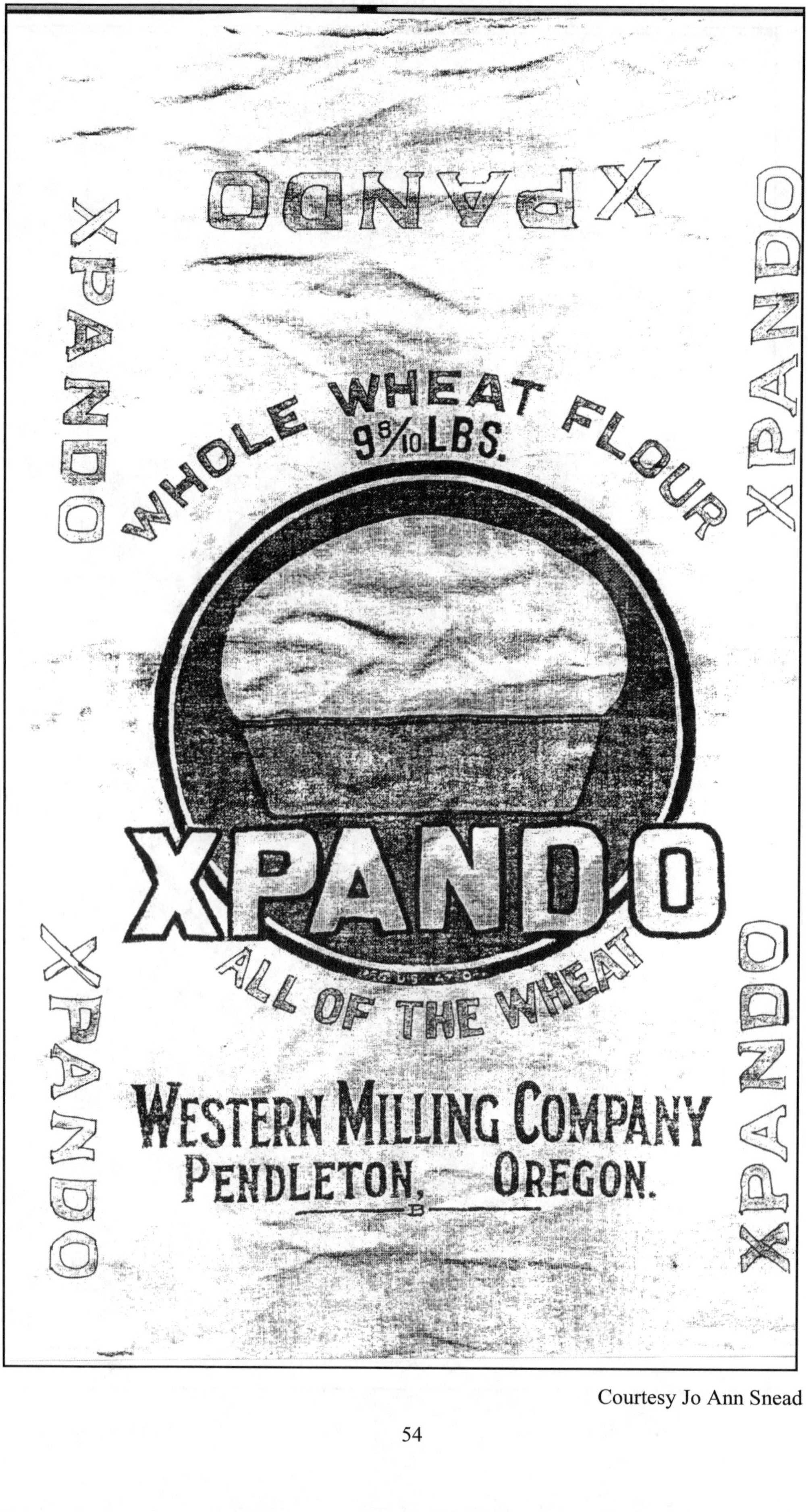

Courtesy Jo Ann Snead

WALLOWA FLOUR MILL

Courtesy Kiddle Family
Given to Marilyn Hulse by Viki Tulley Goodner

This picture was probably taken before 1930. House on the right side was the home of Bruce Cox and later Ted and Bertie Prince. In the year 2000, it is the home of the Tim Burnettes.

The road in front of the flour mill is now known as Highway 82. It had not been paved when this picture was taken.

I am glad to have found photos of the old mills and saved the history of the lofty and stately buildings. When I looked at their pictures the thought crossed my mind, why were they all built so tall? The following information and grist mill diagrams should help explain a little about the operation. Gravity feed was necessary and gristmill builders took full advantage of it.

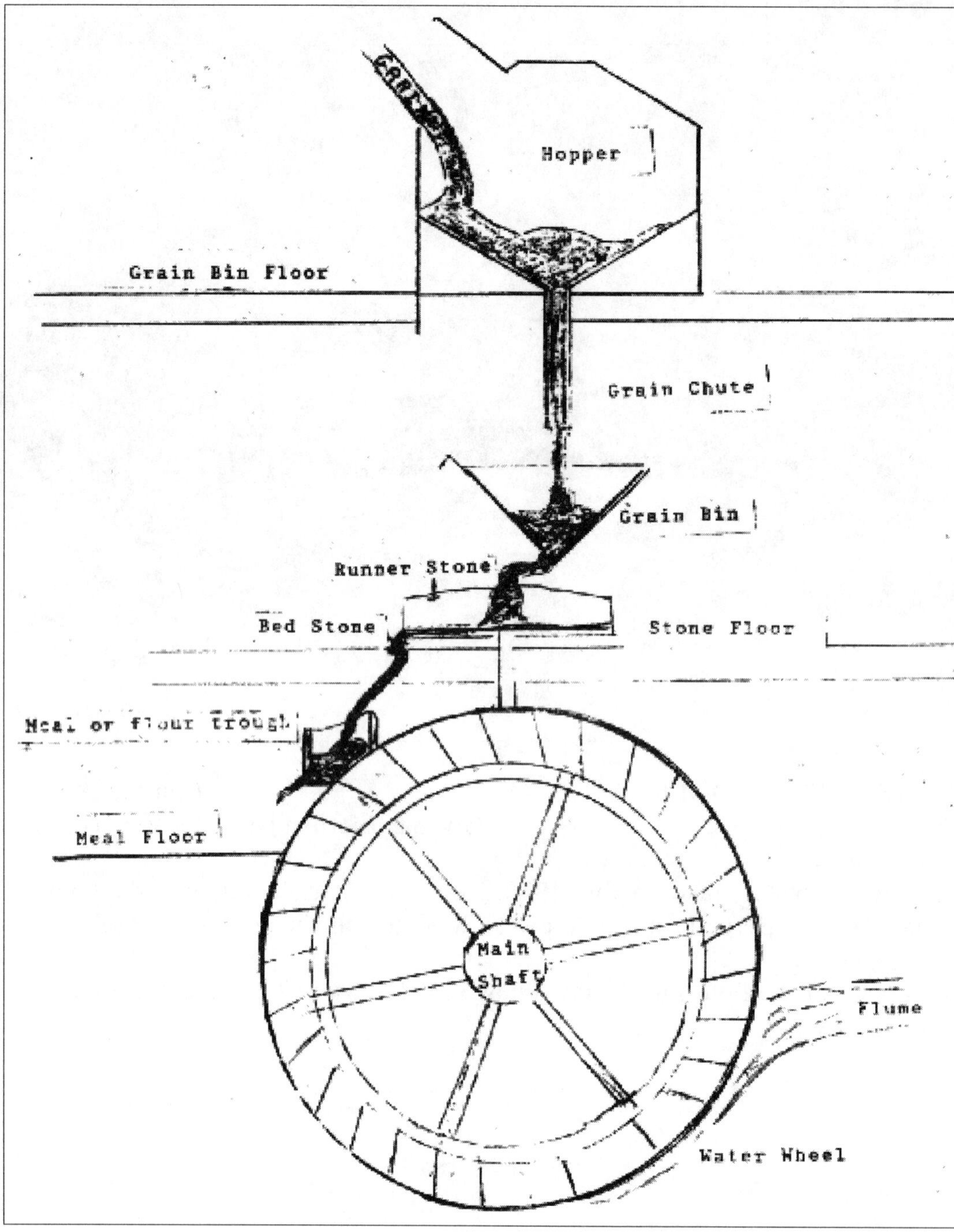

This is a brief sketch of the milling process, as used by the earlier mills.

RURAL AND SMALL TOWN FLOUR MILLS:

By Jack Evans

Windmills and water mills are familiar to us from pictures dating back to the Middle Ages, and their history goes back much farther than that. They changed very little for hundreds of years. Windmills were common in areas with strong prevailing winds, usually where river currents were sluggish, as in the low countries, or where water was scarce; those hunted by Don Quixote were on the Spanish plains.

The windmills were strong, relatively simple machines, and surprisingly sophisticated for their time. The mill was attached to its base by a central pivot, on which it could be turned to catch the wind and then locked into position. The sails could be feathered to avoid buffeting when the mill was not in use. The building's central mechanism, which turned the millstone, was linked to the main shaft by a system of massive wooden gears. These could be engaged, or disengaged, by a simple lever-operated clutch. Millstones were circular and the upper (or runner) turned on the lower (or bedstone). The bedstone had radiating grooves cut in its upper surface. Grain was fed into the central hole of the upper stone (the eye) by gravity and as it was crushed and ground it was forced along the grooves and into a box that surrounded the stones. From here it was gathered into bags and transferred to wagon, cart or storeroom.

These mills were quite efficient for their time. Their only real disadvantages were their limited production capacity and their inability to function during calm weather.

Water mills were similar in principle to windmills but differed in that the mill was stationary and located on a riverbank where the current had enough strength to turn its wheel effectively. Three kinds of wheels were used, of which the overshot wheel was most efficient; this is the kind usually seen in old pictures. It required enough drop in the streambed to allow an overhead flume to be brought in from upstream--usually at a height of about ten feet. Water from this source fell on the paddles, causing the wheel to turn. Where the stream was rapid and an overhead flume was not feasible, the undershot wheel was employed. Water from a ditch or ground-level flume struck the paddles at the bottom to turn the wheel. The breastshot wheel was midway between the other two and very inefficient. The (above) accompanying sketch is of the simplest kind of early water mill; this one used the undershot wheel.

Water mills were the most common type in America until after the Civil War. Their first real changes in centuries came with the Industrial Revolution. By the mid-Nineteenth Century wooden gearing was being replaced by iron and the mills were performing increasingly complex operations. Although the exterior wheel could be raised out of the water when not in use, it did have some disadvantages. It was vulnerable to flood and storm damage, and it could be immobilized by ice.

The growing use of steam provided a reliable source of power for many mills. It was relatively independent of restrictions imposed by weather and seasonal changes, and in winter it gave warmth as well as relief from the labor of chopping ice away from the wheel. It did have some disadvantages of its own: fuel had to be cut and stored, the fire had to be stoked, the water level in the boiler monitored, and the person who maintained it required special training. The engine room also had to be separate from the mill; the air in a flour mill was often charged with fine inflammable dust and one spark could cause a disastrous explosion.

By 1882, when the accompanying sectional drawing of a typical flour mill was published, the small mill had reached a comparatively high level of efficiency. This one is a burr-mill, as those with stones were called, (the first gristmill in Oregon was equipped with burrs and was built at Forest Grove in 1844) and it can be set up for a variety of operations. The stones are now standardized at about four feet in diameter; they and the other equipment are operated by a system of bevel gears in the basement. Grain is dumped into a hopper beneath the ground floor, from which it is raised by elevators (usually endless belts or chains with scoops, running in hollow wood columns) to the attic. Here it falls into bins. From each of these it is led by gravity down a funnel to the eye of its particular stone. The ground meal is then raised by another elevator to the second floor, where it falls into a large sifter assembly. Here it is passed through successively finer screens, as these are shaken, and finally through cloth. At each stage fans remove the chaff and the meal may be reground to refine it further. The final product is then begged and is ready for market. The whole system is largely automated and can be carried on by one miller and one or two assistants.

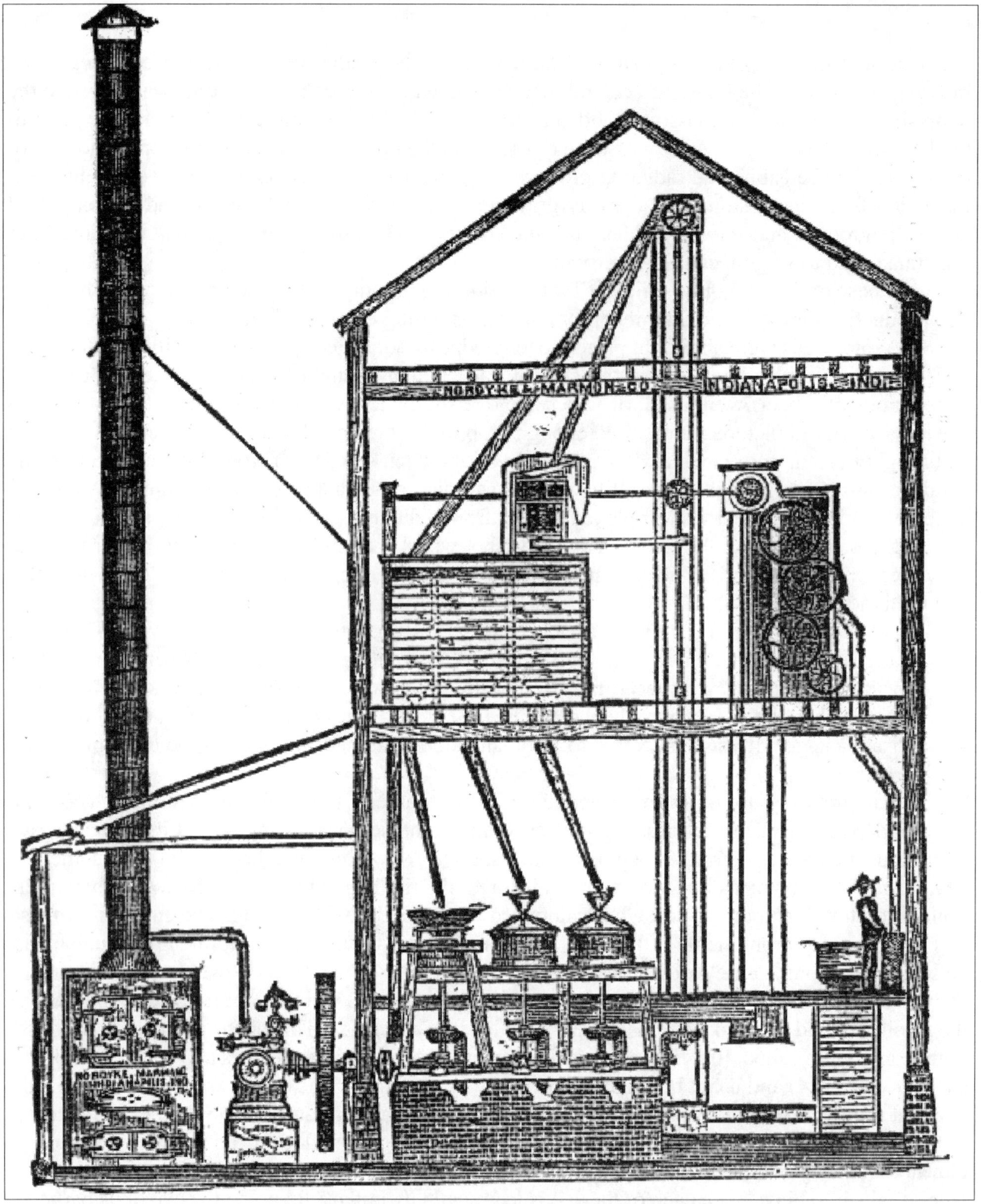

SECTIONAL VIEW OF A FLOUR MILL

From *Moore's Universal Assistant and Complete Machanic (1882)*

The above cut, from the *Mill Stone*, published by well known mill furnishing firm of Nordyke & Marmon Co., of Indianapolis, Ind., affords a sectional view of a complete flour mill with a three run outfit, two runs for wheat, and one run for corn.

There were two final evolutionary steps in small flour mills, both of which greatly increased their efficiency. One was the change from burrs to rollers, in which process the grain was crushed between heavy steel rollers and high volume production was made possible.

The stone burrs, in addition to limited capacity, had certain other disadvantages. The bedstone had to be kept true and the grooves recut occasionally, and this required a skilled artisan (the dosser) to dress the stones. Modern stones were made of granite, which wears slowly and produces little grit; in earlier times and other places, granite was not always available and millers had to settle for the hardest stone they could get. This meant fine grit in the flour, which contributed greatly to dental wear.

The second great improvement was the final refinement in water power. This was the turbine. It required a mill race or impoundment, but there was no longer any mechanism visible outside the mill. A watertight box, about ten feet square and ten to fifteen feet deep, had a platform near the bottom in which was a circular opening. The turbine rotor, resembling a multi-bladed propeller or fan, turned horizontally in the opening. Its axle was vertical, the upper end geared to machinery in the mill, and the lower resting on an underwater bearing. These bearings were made of lignum vitae, an extremely dense tropical hardwood from South America. It is oily and very hard, and it can support such heavy weights under water for an impressive length of time.

The turbine could be disengaged for maintenance or repair and the water diverted from it over the spillway.

This system worked well under most conditions, as it was not vulnerable to ice and required only the weight of a column of water to make it run. The location of the shaft was visible only as a row of angled timbers intended to keep debris (or people and animals) from falling in, and was sometimes marked by small whirlpools on the surface. A further advantage of the turbine was that it was sufficiently powerful to run a dynamo, making an electric lighting system possible. This was far safer than the kerosene lanterns previously in use and of course provided much better illumination. In time some mills converted all their machinery to use electric motors and used the turbine only for generation of power.

The mill buildings were all extremely well built in order to withstand the weight and vibration of heavy machinery, and the millwrights who built them were practical engineers of a high order. Their art has largely died with them.

Production of these mills, in their improved form, ran around 125 barrels of flour per day. Tony Flagg, of Pendleton Flour Mills, has provided the following clarification: "A barrel of flour weighs 200 pounds. The diameter at the head of the barrel is 17 inches, at the bulge it is 20 inches, and the depth is 26 inches." Whole wheat contains all the parts of the wheat kernel, including the germ, also the inner portion from which white flour is made. Whole wheat, or graham, flour was named for a man called Sylvester Graham who preferred it.

Rodents were a serious problem in all the old mills. A crew of cats was maintained to control them, and this was the most efficient method found.

By World War I railway service had become so universal that the huge mills in urban areas, particularly in the midwest, could produce flour in enormous volume, ship it all over the country, and still undersell the local miller. The little mills struggled on, sometimes surviving for a time by concentrating production on feed, but they could no longer compete. Those that remained were idled by the Depression and after World War II the last of them disappeared. The era of self-sufficient small towns, each with its own major resources, had vanished forever. The local flour mills vanished with it.

FLOUR MILLING TODAY

In spite of the enormous scale of such operations today, flour production is not greatly different as the present time. The following brief description is taken from a recent edition of *World Book Encyclopedia*:

"The method by which wheat is made into flour is called milling. The white inner part, or endosperm, of the wheat kernel is separated from the outer, or bran, layers and the germ by many grindings and siftings.

"The wheat is first carried to the top of the mill in elevators. Then it is passed through cleaning and scoring machines to remove foreign substances and dirt. The wheat is then moistened, or tempered, so that the dark outer layer will not break and discolor the flour. After cleaning, the wheat is passed between great rollers to crush the grain. The broken pieces are then shaken in sieves called bolters to separate the flour from the broken bits of grain and the coarser particles. Then the crushing and bolting are repeated until as much as possible of the flour has been separated from the bran. After the first grinding the crushed wheat is separated into first-break flour, middlings, and bran. After each grinding the middlings are purified by removing the bits by means of air currents. The purified middlings are then ground fine to form white flour."

FURTHER READING

If you would like to learn more about gristmills, these two books are both excellent:

A Brief History of Milling, by Charles Howell and Allen Keller

Eastern Washington's Vanished Gristmills and the Men Who Ran Them, by Florence E. Sherfey

If anyone has any more pictures or information of the mills of Wallowa County, Oregon, it would be great to add that material to this story. I would especially like to see a flour sack from Lostine and Enterprise. Call 541-886-3002 or write Irene Barklow, 504 Donald Street, Wallowa, OR 97885.

To order this book "The Forgotten Grist and Flour Mills of Wallowa County, Oregon", write to Enchantment Publishing Of Oregon, Irene Barklow 504 Donald St. Wallowa, Oregon 97885.
Phone 541 886 3002 Or email: barklow33@hotmail.com

INDEX

Irene Locke Barklow was raised near Bolivar, Missouri. Her family moved frequently between Missouri and California in her early years. She attended the Southwest Missouri State College at Springfield, Missouri and in 1981 received credits from Eastern Oregon State College in La Grande, Oregon. Irene came to the Wallowa Valley in 1945 and married Kenneth Barklow in 1946. They have one daughter, Christine, married to Roark Schwanenberg. Christopher and Deena Tsiatsos and Margo Schwanenberg are their Grandchildren. Irene was postmaster at Lostine, Oregon for twenty-five years and has maintained an active interest in the history of northeastern Oregon. In 1965 she and Geneva McArtor started the Lostine Fabric Shop. In six years they had five shops in the Northwest. These were all sold in 1972. It was a pleasure for Irene to be an early board member of the Wallowa County Museum. As Co-Chairman of the Lostine Bicentennial Committee Irene suggested, and co-chaired the first Lostine Flea Market. In 1983 she published her first book, *The Old and The New* of the forty five Post Offices and Communities of Wallowa County; and in 1987 *From Trails to Rails:* The fifty two Post Offices and Communities, Stage Stops and Wagon Roads of Union County Oregon. The largest effort was *School Days in the Wallowas* in 1992, it tells about the ninety-one schools and Communities of Wallowa County, Oregon.